THE NO-WASTE VEGETABLE COOKBOOK

THE NO-WASTE VEGETABLE COOKBOOK

RECIPES AND TECHNIQUES FOR WHOLE PLANT COOKING

LINDA LY

PHOTOGRAPHY BY WILL TAYLOR

HARVARD
COMMON
PRESS

Brimming with creative inspiration, how-to projects, and useful information to enrich your everyday life, Quarto Knows is a favorite destination for those pursuing their interests and passions. Visit our site and dig deeper with our books into your area of interest: Quarto Creates, Quarto Cooks, Quarto Homes, Quarto Lives, Quarto Drives, Quarto Explores, Quarto Gifts, or Quarto Kids.

Inspiring | Educating | Creating | Entertaining

© 2020 Quarto Publishing Group USA Inc.
Text © 2015, 2020 Linda Ly

First Published in 2020 by The Harvard Common Press, an imprint of The Quarto Group, 100 Cummings Center, Suite 265-D, Beverly, MA 01915, USA.
T (978) 282-9590 F (978) 283-2742 QuartoKnows.com

The Harvard Common Press titles are also available at discount for retail, wholesale, promotional, and bulk purchase. For details, contact the Special Sales Manager by email at specialsales@quarto.com or by mail at The Quarto Group, Attn: Special Sales Manager, 100 Cummings Center, Suite 265-D, Beverly, MA 01915, USA.

24 23 22 21 20   1 2 3 4 5

ISBN: 978-1-55832-997-3

Digital edition published in 2020

The content in this book was previously published in *The CSA Cookbook*, by Linda Ly (Voyageur Press 2015).

Previously cataloged under the following Library of Congress Control Number: 2014952944

Design: Amy Sly
Cover Image: Will Taylor
Page Layout: Sporto
Photography: Will Taylor

Printed in China

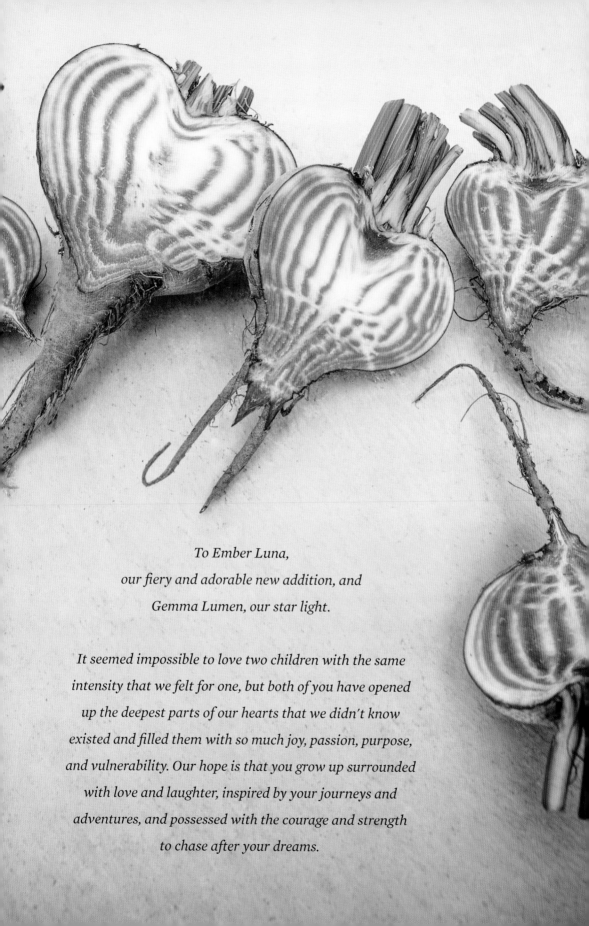

To Ember Luna,
our fiery and adorable new addition, and
Gemma Lumen, our star light.

It seemed impossible to love two children with the same
intensity that we felt for one, but both of you have opened
up the deepest parts of our hearts that we didn't know
existed and filled them with so much joy, passion, purpose,
and vulnerability. Our hope is that you grow up surrounded
with love and laughter, inspired by your journeys and
adventures, and possessed with the courage and strength
to chase after your dreams.

CONTENTS

PREFACE

Growing up in an Asian household, I've seen and eaten my share of somewhat strange foods: pepper leaves, pea shoots, kale flowers, and sweet potato vines, not to mention the nose-to-tail portions of nearly every animal that appeared on my plate. Of course, they didn't seem strange to me at the time; they were just what's for dinner. Most of our meals were vegetable-centric, due in part to a sense of thrift and the food culture in my family's homeland. In their belief that no food should ever be wasted ("Because there are children starving in Vietnam!" they'd say), my family fully utilized every part of the plant they brought home from the market.

This was all before the back-to-the-land movement in urban backyards and the farm-to-table trend in upscale restaurants. We were eco-friendly before it was even a "thing," and it's not that we were especially ahead of our time. My parents lived through war-era Vietnam, where supplies were scarce. Since refrigeration was a luxury, they shopped as needed, every day, from one of the many farm stalls down the street.

Food was always fresh, because that was the only way my parents knew how to cook and eat. They couldn't afford a microwave in the early days of life in America, so they cooked the old-fashioned way: on a stove, with a bevy of pots and pans. Even after a long day at work, my mom and dad would dutifully prepare dinner as a way to wind down and catch up with each other; many a chatter and debate was had in the kitchen over a sink full of salad greens.

I envied our neighbors who served up quick, tidy meals made from cans and boxes. Meanwhile, we were meticulously rinsing rice, washing herbs, chopping vegetables, and steaming whole fish, head and all. Everyone took part in the prepping as a nightly ritual, and nothing was wasted.

Coming from a culture where food was a celebration of life and meat was a special-occasion splurge, my family sat down every night to home-cooked meals that consisted of one part meat and three parts vegetables, long before former First Lady Michelle Obama ever promoted a new food "pyramid" that encouraged something similar.

I was taught to eat everything off my plate (because there are children starving in

Vietnam, of course) and taught that all parts of the plant were one and the same—that the roots, stems, leaves, flowers, and seeds were equally precious. Whenever I tried to wiggle my way out of eating, say, Chinese broccoli stalks, my mom would remind me that *[insert random palnt part here]* was *bổ*, or nutritious. According to her, every vegetable was good for treating or curing an ailment, whether it be the common cold or high cholesterol. (It wasn't until decades later that I finally realized Mom was right—food *is* medicine.)

I can't say I really appreciated any of that as a kid, which is probably true of a lot of first-generation Asian-American kids. I rebelled against the horror that was homemade food and complained about the time it took to prepare a meal when pizza was just a phone call away. I even tried to trade lunches at school—but who would want a *bánh mì*, a Vietnamese baguette filled with pâté, pickles, and herbs, over a PB&J made with Wonder Bread, Skippy, and Smucker's?

In spite of it all, fresh whole food was what I'd always known. It was familiar and, well, it was made for me and made with love. I ate whatever my parents put on the table—things that other people thought were exotic, but for me was just another trip down the Asian grocery aisles. Bags of pea sprouts and bundles of budding garlic chives were as ho-hum as a head of cabbage.

It wasn't until after college, when I started paying more attention to food and learning how to cook for myself, that I realized how much of it is often wasted in the American kitchen. Wandering the produce aisle of the local grocery store, I would see sad bunches of greens attached to beets and neatly bundled leeks with dark green leaves that looked like an afterthought. Many of my friends didn't even know you could eat radish tops, let alone carrot tops. It seemed like half the vegetables you purchased went straight to the trash or, at best, made for some rather expensive compost. For someone whose childhood dinner rotation included stems and shoots and other "trash," it was startling and eye-opening.

As soon as I started cooking, other people's discards were destined for my dinner plate. I became quite adept in scrap artistry, taking the odds and ends of my vegetables and turning them into full-fledged "kitchen pantry" meals, meals that came together with whatever ingredients I had on hand in my tiny apartment on an even tinier budget. I was a fan of vegetables that offered twofers (or more), such as crisp, raw beets with tender leaves or pumpkins that provided sustenance through their seeds, flesh, flowers, and leaves. Cooking this way was less wasteful, for sure, but also more interesting. Flavor can be found in unexpected places.

FAST FORWARD TO THE SUMMER OF 2010

I moved into a little house with a big, big backyard on the Southern California coast—virtually unheard of in a place where land is a commodity. Until that year, I never truly had a space to call my own. All the communal rooftops and fire escapes in the apartments and lofts I lived in for nearly a decade were just that—urban environments more conducive to xeriscaping than edible planting. I was intrigued by gardening but also discouraged, as the small spaces and limited light did little to support the potted plants I tried to grow.

But in this new house, the previous owners had started an organic vegetable garden right off the kitchen, an impressive array of raised beds set on a terraced slope with southern exposure. When I moved in, smack dab in the middle of summer, the garden was already going off with tomatoes, peppers, beans, and squash of all kinds.

Truth? I was intimidated by this garden and didn't step foot inside it for the first couple of weeks. After a dismal track record of killing many an herb on my windowsill, I was worried I'd kill this garden, too. But as I began to harvest one, then two, then four zucchini a day from plants that thrived (or survived?) under my care, I realized that even thumbs as black as mine could be converted. As I like to say now, there are no black thumbs, only green thumbs in training.

That fall, I harvested my first crop from seeds I'd started earlier in the season: Japanese radishes, whose sizeable, spicy leaves were as delectable as the roots. There was no question that I would eat the leaves, but rather, *how* would I eat the leaves? (From that harvest came one of my favorite early recipes, Portuguese Sausage and Radish Green Soup, page 143.)

CURIOSITY IN THE GARDEN, CREATIVITY IN THE KITCHEN

Over the years I've started, tended, harvested—and yes, even killed—hundreds of varieties of vegetables and herbs, mostly heirlooms and mostly things you would never see in a supermarket or farmers' market. I have an affinity for unusual, gnarled, dimpled, dappled, scarred, and otherwise unconventional varieties—what others might view as flawed, but I consider beautiful.

I've cultivated countless vegetable plots, container gardens, herb gardens, flower beds, and fruit trees. Growing my own food has made me more inclined to use all the edible parts of plants from my garden. Once you realize how much work goes into them—the seeding, watering, weeding, feeding, trimming, and harvesting—you don't want to waste them. Had you told me, a recovering city girl, that one day I'd be an urban farmer, sharing my love of getting dirty and cooking clean on a blog called *Garden Betty*, I would've told you to get out of town.

But here I am. I've taken my love of edible gardening from the year-round growing climate of Southern California to the more challenging seasonal conditions of central Oregon, where I continue to grow as much of my own food as possible, for both pleasure and health.

Among my favorites are the vegetables normally neglected and seldom used, from the shield-shaped leaves of beans that climb up a pole to the big, beautiful leaves that grow around a head of broccoli. These are leaves that shoppers never see and recipes never call for (so you'll be happy to find some on pages 79 and 90, respectively).

I think of myself as a garden foodie, someone whose best meals are pulled from the soil just hours before, prepared at home, and enjoyed without fuss. When I'm asked what my most memorable meals have been, through all of my travels around the world, the answer is not fine dining in Manhattan or street food in Bangkok (though those come very close). It's a farm-to-table meal from my own modern homestead, simple and fresh as can be.

You see, I have a thing for peasant food—the classic comfort dishes that bring family and friends together. I believe that good food doesn't need a lot of tinkering to make a meal special. Good food should be inventive yet accessible, nourishing yet inexpensive.

It's farm food, the exact opposite of fancy food, and it's a cuisine in which creativity and resourcefulness does not equal frugality and blandness. It's understanding that a vegetable begins with the sprouts and does not end until the tubers, vines, leaves, flowers, fruits, and seeds have given their all.

Although the idea of eating the tops and tails of vegetables may seem radical or exotic, I assure you it's not as bizarre as it sounds. All over the world, other cultures consume these parts as their everyday food. Countries that are not as meat-centric as the United States tend to be more creative in cooking with them, as vegetables are less expensive and more abundant. In the Philippines, pepper leaves are the main ingredient in *tinola*, a traditional chicken soup. In Thailand, squash shoots are sautéed in coconut milk and curry. In Lebanon, chard stalk hummus is as commonplace as chickpea hummus.

Like nose-to-tail butchery, whole vegetable cooking is a beast of its own. While many people are becoming more in tune with where their food comes from, they still feel stumped beyond the conventional uses for their vegetables. Kitchen scraps seem solely relegated to soup stock, as if they're no good. But every part of the plant, even the so-called scraps, imparts a different flavor and texture to a dish. By learning about the often underutilized parts of plants, like the green tops of leeks or the yellow buds on kale, you'll open yourself to a whole new range of vegetables.

INTRODUCTION

I'm sure there has been many a time when you've eyed a beautiful bulb of fennel with a full head of fronds at the farmers' market and thought, "What do I do with that?" And maybe you've fed bunches of carrot tops to your backyard chickens or chucked the dark green ends of leeks in your compost pile. Those days are no more.

My intent with this book is to show you the boundless possibilities of produce—applying the nose-to-tail approach to vegetables in what I like to call top-to-tail (or seed-to-root?) cooking. Although everyone knows you can throw a radish in a salad, how about a radish in a hot pan, braised with butter and served with wilted greens? (Butter-Braised Radishes and Radish Greens with Farro, page 145.) If you find some strange-looking bulbs at a farmstand that resemble Sputnik 1, I have recipes for eating them raw (Kohlrabi and Carrot Slaw, page 107), tossing and roasting them (Kohlrabi Home Fries with Thyme Aioli, page 108), and sautéing the leaves (Kohlrabi Green and Wild Mushroom Ragoût with Polenta, page 111). If you grow tomatoes at home, have you ever wondered what you could do with those verdant vines that sometimes grow up to 10 feet (3 m) long? There's a sauce for that. (Spicy Minty Tomato Sauce Infused with Tomato Leaves, page 34.)

I've always considered cooking as an art, rather than an exact science. That said, I tend to assume that every recipe is variable when it comes to cooking times and temperatures. Kitchens differ so widely with the pans we use and the stoves we cook on, no time or temperature is ever set in stone. While these numbers are provided in every recipe, I also give sensory cues throughout the cooking process so you can gauge the doneness of the ingredients. For me, the nose and the eyes are the most accurate indicators in a kitchen.

Many recipes can easily adapt to the ingredients you have on hand as well. Leafy greens with similar textures can often be used interchangeably; if you don't grow broccoli greens, you can substitute collard greens, and if you're out of chard, you can use beet greens. Carrot tops can take the place of parsley in a pinch, and vice versa. Mix and match the pickling brines with whatever root vegetables you have in the kitchen, and if it's green, it can almost always be pounded into a delectable sauce (see A Primer on Pesto, page 29). Food should be fun, and I encourage you to experiment with the recipes and make them your own. The more intuitive you are, the better the dish will turn out.

The No-Waste Vegetable Cookbook was written not as a manifesto on food waste or food security, but as a celebration of the art of using the whole vegetable in your day-to-day cooking. I hope it will embolden you to look at food in a different way, be spontaneous in the kitchen, and feel inspired by your backyard harvest, farmers' market basket, or everyday grocery run.

THE BASICS

PANTRY AND KITCHEN NOTES

For me, a recipe might begin with a basket of peas from my garden or a bundle of greens I've never tried before from a farmers' market. With no menu in mind, I always look to my pantry for cooking inspiration—a pantry that favors classic, economical, and easily accessible ingredients over the diet du jour. Nothing is more frustrating than finding a recipe you're really excited about, only to realize you have to make a trip to a specialty store to source an obscure spice. Most of the ingredients in this book are readily available at your local grocer, and some of the less common items can be found in the ethnic food aisles of a well-stocked supermarket. (That said, I always advocate supporting small businesses and buying local or organic whenever possible, but I'd guess that anyone reading this book is already in that mindset or striving to be.)

THOUGHTS ON ORGANIC

In truth, I'm not a stickler when it comes to organic purchases. There are plenty of small, honest food producers who abide by organic principles but cannot afford organic certification. I'm actually more dubious of food with health claims splattered on the packaging, "natural" foods that really aren't, and food products that latch onto food trends. If you're trying to shop with an eco-conscious mind, just do the best you can with the resources you have, and remember that organic labels are not the be all and end all. Sometimes we analyze food a little too much and take all the joy out of eating!

A WORD ON SALT

The recipes throughout this book call for specific amounts of salt during prepping or cooking to disgorge or enhance the ingredients. Since salting is subjective, feel free to add more at the end of the recipe if you feel the dish still needs it. For recipes that require chicken or beef broth, it's assumed that your broth is already well seasoned; if you use virgin stock, you may need to add some salt to taste. Likewise, if you use unsalted butter, recipes that call for ¼ cup (55 g) of butter or more may taste too bland without additional salting.

In my day-to-day cooking, I like to use a medium-grained Himalayan pink salt or flaky sea salt; I find that the coarser grains give a more intriguing and layered taste in each bite, which heightens the flavor of the whole dish.

SIZING STANDARDS

Unless quantities are specified by weight or volume, figure all vegetables (including onion and garlic) to be medium or average in size. "Medium" can be a somewhat ambiguous term, but if I feel a little more or a little less won't make or break the dish, I'll simply call for a basic whole quantity in my recipes.

TO PEEL OR NOT TO PEEL

With few exceptions, I never peel my vegetables. I peel onion and garlic, of course, as well as beets and kohlrabi, whose skins are too gritty and tough for my taste. But I never peel ginger, since grating or mincing the root makes the small bits of papery skin undetectable in a dish. I also never peel carrots (which I think are more beautiful with their skins on) or russet potatoes (unless I'm making super creamy mashed potatoes).

With fresh vegetables, especially those organically grown, peeling tends to be more of an aesthetic preference. There's really no need to produce all that waste when the skins are perfectly edible, nutritious, and delicious. Just make sure you wash and scrub them well with a vegetable brush before using.

MEASURING FLOUR

In this book, flour is measured without sifting first. I use the "scoop and sweep" method: scoop a heaping mound of flour with your measuring cup, then sweep a straightedge across the cup to level it. Do not tamp down the flour or tap the cup on a surface to settle it. If your flour has been compacted at the bottom of a bag or canister, lightly fluff it up with a fork before scooping.

TOASTING HOW-TO

Because toasting times for nuts vary widely depending on the size of the nuts and the accuracy of the oven, I prefer to use the dry pan method. For toasting small quantities of nuts, it's the surest way to keep them from browning (or burning) too quickly.

Heat a small skillet over medium heat. (No fat is needed; you want the skillet dry.) Spread the nuts across the surface and toast, stirring occasionally, until they release a rich and nutty fragrance. You want a golden brown or slightly darker color overall but never any black spots.

INGREDIENTS

A large part of my day is spent cooking in a small kitchen with a tiny pantry (if you can even call an under-the-counter cabinet a pantry), so the ingredients and tools I'm able to squeeze into that space are ones I rely on almost daily. The brands suggested are solely that—suggestions. They're everyday brands I personally use and trust, and the tools I recommend are what I consider indispensable, making many appearances throughout this book.

SALT

For consistency in taste and measurement throughout the book, I use Morton Coarse Kosher Salt. If you don't use this same brand, think of it merely as a reference point when seasoning with your own salt. If you use finer-grained crystals, start by halving the amount of salt called for in the recipes and add more to taste.

SUGAR

Unless otherwise noted, all sugar is C&H Pure Cane Sugar (a white granulated sugar).

ALL-PURPOSE FLOUR

Recipes in this book that call for flour use Gold Medal Unbleached All-Purpose Flour, a middle-of-the-range flour with 10.5 percent gluten. Keep in mind that different brands of all-purpose flours have varying gluten content, especially across regions, so use your best judgment if you feel your dough needs more or less water to hold its shape.

BUTTER

For all of my savory cooking, I use salted butter. I rotate among common supermarket brands, but the average stick (½ cup [112 g]) of salted butter that I buy contains around ¼ teaspoon salt. If you prefer unsalted butter, you may need to add or adjust the amount of salt called for in the recipe, especially if it requires ¼ cup (55 g) of butter or more (anything less and you likely won't taste a difference).

MILK

Unless otherwise noted, milk is always whole, and other dairy products used (such as sour cream and ricotta cheese) are full fat.

CHEESE

I highly recommend grating or shredding your own blocks of cheese, rather than buying bags of preshredded cheese. Preshredded cheese contains a powdered additive called cellulose, which prevents the cheese from clumping together. This same additive also makes it more difficult to melt the cheese evenly in recipes that call for melting.

EGGS

Egg size is always large. Whenever possible, I encourage you to buy eggs from pasture-raised hens that are free to forage for bugs, grasses, weeds, and seeds. Studies have shown that aside from being ethically produced, eggs from pastured hens are nutritionally superior to standard supermarket eggs, containing less cholesterol and more vitamin A, vitamin E, and omega-3 fatty acids.

OLIVE OIL

For day-to-day cooking, I use an inexpensive, light, and fruity extra-virgin olive oil. For drizzling over dishes or whisking into dressings, I prefer a rich and robust extra-virgin olive oil. Supermarket brands that I regularly buy are California Olive Ranch and Kirkland Signature (from Costco), as the oils are always fresh and reliable. If you're feeling overwhelmed by the selection at the store, choose a cold-pressed extra-virgin olive oil sold in a dark bottle (which helps prevent oxidation) and let your nose and taste buds guide you. The best olive oil is one that you personally like, and not necessarily the most expensive one.

SUNFLOWER OIL

When a neutral flavor or high smoke point is needed for cooking, I call for sunflower oil. There are plenty of other good choices on the market, so feel free to swap in cold-pressed or expeller-pressed safflower, grapeseed, or avocado oil when flavor or smoke point is a concern.

SESAME OIL

Asian sesame oil is used throughout this book, and the brand you use should be roasted, heady in fragrance, and dark in color. A little goes a long way for this smoky oil. I use Kadoya Pure Sesame Oil, a Japanese brand found in most Asian markets, as well as the ethnic food aisles of well-stocked supermarkets. When buying a different brand, look for a dark-colored toasted sesame oil.

SOY SAUCE

In all my cooking I use Chinese soy sauce, which has a more savory soybean flavor than Japanese soy sauce (shoyu, which is sweeter). Chinese soy sauce is often sold as light (not to be confused with the low-sodium versions sometimes labeled as "light") or dark. For day-to-day cooking, light soy sauce (also called regular soy sauce, thin soy sauce, or fresh soy sauce) is preferred for its cleaner, sharper flavor. If you cannot tolerate wheat, look for tamari, a Japanese soy sauce similar in style to traditional Chinese soy. Tamari is made with more soybeans and little to no wheat; not all brands are wheat free, however, so be sure to check the label before you buy.

My pantry staple is Wan Ja Shan, a rich and aromatic soy sauce available in well-stocked Chinese and Vietnamese markets. Wan Ja Shan also has an American-brewed line out of the Hudson Valley in New York, and its organic soy sauce and organic gluten-free tamari are quite good. The American Wan Ja Shan line is carried by select Whole Foods Market and Sprouts Farmers Market locations, as well as some specialty grocers and better supermarkets. If you can't find this brand locally, look for a naturally brewed soy sauce with water, soybeans, wheat, and salt as its only ingredients. Many brands contain unnecessary additives.

FISH SAUCE

Also called *nước mắm*, fish sauce is the amber liquid extracted from fermentation of anchovies. I always use Vietnamese fish sauce in recipes that call for it, as it tends to be less salty and more complex in flavor than the common Thai and Filipino varieties. Thai fish sauces dominate the shelves at most Asian markets, and some will even have Vietnamese writing on their labels. But if you want proper Vietnamese fish sauce, look for a brand fermented and bottled in Phú Quốc, the famous fish sauce–producing island in Vietnam. Ensure that anchovy and salt are the primary ingredients.

At home, I use Red Boat, a *Phú Quốc* fish sauce with wild-caught black anchovies and sea salt as its only ingredients. It's a pure Vietnamese product but has a Western-friendly vibe with labeling in English. Look for it in select Asian markets, health food markets, and Whole Foods Market locations.

CHICKEN BROTH

I almost always use chicken broth in my recipes, but you can substitute vegetable broth if desired. In an ideal world, I like to make my own long-simmered, health-boosting bone broth. But truth is, I don't make my own chicken stock too often, as I have no room to refrigerate or freeze the amount I would go through in any given week. (I'm a big soup lover.) Instead, I rely on the organic soup bases from Better Than Bouillon and the soup bases from Penzeys Spices, which I like equally. If you prefer canned or boxed broth, bouillon cubes, or other concentrated pastes, check the label and ensure that meat (or a vegetable), not salt, is listed as the first ingredient.

WINE

When I call for white wine in my recipes, any midrange, dry white wine will do. I usually reach for a Pinot Grigio or Sauvignon Blanc. For a dry red wine, my standbys are Cabernet Sauvignon, Merlot, or whatever wine I happen to be drinking that evening. (And if you're wondering, Malbec is one of my favorites!) Never use a bottle labeled as "cooking wine," as it's bottom-of-the-barrel wine to which salt and food coloring have been added. Using wine you like to drink will always taste better.

TOOLS

POTS AND PANS

For the recipes in this book, saucepans are referenced as small (1 quart [946 ml]), medium (2 quarts [1.9 L]), or large (4 quarts [3.8 L]). Pots are anything larger than that, and a wide, heavy pot is a 6-quart (5.7 L) or larger pot that's wider than it is deep (such as a Dutch oven). Skillets are small (8 inches [20 cm]), medium (10 inches [25.5 cm]), or large (12 inches [30 cm]). When I call for an ovenproof skillet, my go-to is a heavy cast-iron skillet. I use a 5-quart (4.7 L) sauté pan when the higher, straighter sides of the pan (as opposed to the shallower, curved sides of a skillet) facilitate cooking. In the few instances where a wok is recommended, choose a multi-clad or heavy-gauge, flat-bottom wok. I prefer well-seasoned cast-iron pans to polytetrafluoroethylene (PTFE)–coated nonstick pans, but they can be used interchangeably (except for stir-frying, in which the high heat causes nonstick coatings to vaporize and release noxious fumes).

KITCHEN SCALE

Sometimes I feel a scale is the unsung hero of a kitchen. While you might never strictly need it, when you do use it, you inevitably become a better cook. Even though I'm not the most finicky or precise cook, I find myself reaching for my Escali scale a couple of times a week. If you're in the market for one, look for a compact digital version that weighs in pounds and ounces as well as kilograms and grams and has a capacity up to 11 pounds (455 g) or 5 kilograms (which is sufficient for most home cooks).

MICROPLANE ZESTER

For zesting citrus peels or grating hard cheeses (not to mention ginger, garlic, cinnamon, or nutmeg), no tool makes an easier go of the task than a Microplane zester. The wandlike tool takes up little space in a drawer, and I find myself using it just as often as my box grater.

MANDOLINE

When a recipe calls for a lot of slicing, a mandoline can cut your prep time in half. Any good mandoline will come with a finger guard, a dial for adjustable thicknesses, and several blade inserts for slicing and shredding. The multipurpose tool not only makes thin, uniform slices, but also julienne strips, French fries, and crinkle cuts.

JULIENNE PEELER

If you own a mandoline with interchangeable blades, you likely won't need a separate julienne peeler. But if you can squeeze only one more tool into a tiny kitchen, this low-profile peeler is perfect for making mounds of zucchini noodles, carrot strips, and shoestring potatoes in no time at all.

FOOD PROCESSOR

For mixing pie dough, making salsa, shredding blocks of cheese, blending pesto and hummus, or breaking down a lot of food at one time, a food processor is a worthy investment. My 11-cup (2.6 L) Cuisinart suits all of my household's cooking needs and makes even the most mundane chopping chore easy breezy. If you don't have the space for a full-size food processor, I suggest a midsize 6- to 8-cup (1.4 to 1.9 L) model, as it's a more versatile workhorse.

IMMERSION BLENDER

An immersion blender is ideal for pureeing soups right in the pot—no more messy transfers and multiple batches between pot and blender. It also makes homemade aioli a cinch (see the sidebar on page 119) and blends pestos and sauces silky smooth.

DON'T SPOIL THEM ROTTEN

Did you know that vegetables are composed primarily of water? They contain high amounts of water in proportion to their weight; even something as solid seeming as zucchini is made up of 95 percent water, and white potatoes—which have the lowest water content—are still 79 percent water. When a vegetable is pulled out of the soil or picked from a plant, depriving it of precious H_2O, the cell walls start to lose moisture and eventually collapse, causing wilting.

The key to preventing vegetables from going limp is to create a breathable barrier between the moist vegetable and the dry air of your fridge; that means creating an environment that's airy and damp, but not stifling and wet. Plastic bags and kitchen towels work wonders for this; I like to reuse produce bags (the same ones they often come in) and repurpose clean rags (from deconstructed T-shirts, bed sheets, or threadbare bath towels), as they're thrifty, take up little space, and can be tucked into any available nook in the fridge. I tend to store all my vegetables this way on the shelves, where I can see them (forgetting what you have is often the first cause of wilted produce).

If you are anti-plastic, you can also roll up your vegetables in flour sack towels or linen tea towels before storing them in your crisper drawers. In general, keep vegetables and fruits in separate drawers, and keep leafy greens in their own drawer if you can. The tender greens are most susceptible to wilting if kept in close proximity to ethylene-emitting produce.

A good rule of thumb for determining how to store a vegetable is to visit the produce section of a supermarket. Vegetables that are kept chilled and damp with overhead misters (such as leafy greens, broccoli, carrots, and scallions) need cold and humidity. Vegetables that are kept dry in the middle of the produce section (such as tomatoes, potatoes, and onions) thrive in the same environment in your kitchen.

If you find that your carrots aren't as crisp and your greens aren't as snappy as you'd like them to be, you can revive them in a sink full of cold water (the colder the better). Within a half hour, they'll perk up again as their moisture is replenished through osmosis.

All of these methods will keep your vegetables fresh for at least a week. There's much variation depending on the quality of your vegetables and the temperature and humidity of your fridge, but if you find that your food spoils faster than you can use it, consider buying only what you need for a couple of days. There's no need to infinitely store vegetables if you're not going to use them when they're at their freshest.

TOMATOES
Tomatoes should never be refrigerated. If the fruit isn't fully ripe yet, the cold temperature will stop the ripening process and leave you with pale, bland tomatoes. If the fruit *is* ripe and you refrigerate it for more than a few days, it can suffer from what agronomists call "chilling injury," which causes the texture to turn mealy. The best bet is to store your tomatoes at room temperature out of direct sunlight.

PEPPERS

Store sweet and hot peppers unwashed in a loosely sealed plastic bag in the fridge.

LEAFY GREENS

To ensure they last, wash all greens before you store them. Fill a sink or wash basin with cold water, then soak and swish the greens around for a few minutes to let any sediment sink to the bottom. Lift the greens with your hands and shake off the excess water. Wrap them in a dry towel, place in a loosely sealed plastic bag, and refrigerate. If you have a salad spinner, give your greens a good spin and refrigerate them right inside the spinner.

BROCCOLI

Leave broccoli unwashed and refrigerate in a loosely sealed plastic bag.

PEAS AND BEANS

Refrigerate unwashed peas and beans in an airtight container. If you find a lot of condensation on the lid, place a dry towel inside to absorb excess moisture. Shell them only when you're ready to use them.

SCALLIONS AND LEEKS

Wrap unwashed scallions and leeks in a damp towel, place in a loosely sealed plastic bag, and refrigerate. If they're cut, seal them tightly in a plastic bag before storing again.

If you don't use scallions often enough before they get slimy, you can store them upright in a glass of water on your kitchen counter, as long as the roots are intact. They'll "regrow" over the course of several weeks. Simply snip off what you need from the top of the stems, and new stems will continue to grow from the base. Change the water periodically if it starts to look murky.

CARROTS, RADISHES, BEETS, AND OTHER ROOT VEGETABLES

Always trim the greens from the roots as soon as you bring them home; leave about an inch (2.5 cm) of stem on the roots. If left attached, the greens will continue to draw moisture from the roots and cause them to dry out (and thereby go limp). Wash and store the greens as you would other leafy greens. (See "Leafy Greens," left)

Leave the roots unwashed, loosely seal in a plastic bag, and refrigerate. If you like the convenience of prewashed vegetables for snacking, gently scrub the roots clean and refrigerate them in containers filled with water. Change the water every couple of days to keep the vegetables fresh.

CUCUMBERS

If your kitchen stays relatively cool, store your cucumbers at room temperature out of direct sunlight and away from ethylene-producing tomatoes and fruits. (See page 26.) Otherwise, wrap them individually in dry towels to absorb excess moisture, place them in a loosely sealed plastic bag, and store them in the fridge.

SUMMER SQUASH, WINTER SQUASH, AND MELONS

Summer squash and melons should be stored in a cool, dry, and dark place, such as a cupboard or counter. Winter squash should be stored the same way, and in ideal conditions, will keep for several months. Unused cut squash and melons should be wrapped in plastic or stored in a closed container in the fridge.

GARLIC, ONIONS, POTATOES, AND SWEET POTATOES

Store in a cool, dry, and dark place in well-ventilated baskets or bins, paper bags, or mesh bags. Refrigerate cut onions in an airtight container or tightly sealed plastic bag, and refrigerate cut potatoes or sweet potatoes in a bowl of water to prevent browning.

HERBS

In general, most herbs do best when they're left unwashed, wrapped in a damp towel, and stored in a plastic bag or closed container. If you like your herbs prewashed and prechopped for adding to meals on a whim, make sure they're thoroughly dried (in a spinner or on a towel), and store them in an airtight container lined with a dry towel to absorb excess moisture.

Soft-stemmed herbs like basil, parsley, and cilantro also do well when they're treated like a bouquet of flowers. Snip the bottom of the stems and place them in a glass of water, out on the counter at room temperature. Change the water every couple of days if it starts to cloud.

HIGH ETHYLENE PRODUCERS

High ethylene producers are fruits and vegetables that give off high amounts of a naturally occurring, odorless, colorless gas and plant hormone called ethylene. Ethylene triggers cells to degrade, turning the leaves droopier, the skin a different color, and the fruit softer and sweeter (a process we know as ripening). To help prevent overripening and spoilage, keep these high ethylene producers away from other fruits and vegetables:

Apple	Cantaloupe	Guava	Nectarine	Pear
Apricot	Cherimoya	Honeydew	Papaya	Plum
Avocado	Feijoa	Kiwi	Passion fruit	Quince
Banana	Fig	Mango	Peach	Tomato

EVERYTHING-BUT-THE-KITCHEN-SINK STOCK

Say the words "kitchen scraps," and most people will automatically associate them with vegetable stock—or as I like to call it, everything-but-the-kitchen-sink stock (because that's exactly what goes in mine).

If you have a fairly vegetable-focused diet, there's no reason to buy vegetables specifically for making stock. All those tops and tails you typically toss in your compost bin are good candidates for your stockpot instead. You can use vegetables that have lost their luster, but are still perfectly edible; limp celery, fennel stems, and overly ripened tomatoes, which at first seem futile in the kitchen, really shine in a long-simmered vegetable stock.

When assessing your scraps, discard any part that's rotten, blemished, or too past its prime to be considered edible. Even though we're using "scraps" for the stock, we want them to be good scraps. Make sure everything is washed and clean, including onion skins if you use them. In a large household, you can collect your scraps in a gallon-size (3.8 L) resealable plastic bag and store it in the crisper bin for up to a week. When it's full, it can make enough stock for a standard soup pot or Dutch oven. Small households can freeze their scraps until they have a few cups' worth to simmer.

A good vegetable base always starts with a mirepoix, the holy trinity of aromatics: onions, carrots, and celery. Each element brings its own flavor to the table, and the sum of the stock becomes greater than its parts. You don't have to use those specific ingredients to create a flavorful stock, however. By using vegetables with similar flavor profiles, and a good variety of them, you can make a well-balanced stock with the odds and ends you *do* have on hand.

In a typical week of cooking at home, you might find yourself with a pile of these common kitchen trimmings. Far from being a comprehensive list, this is just a starting point for looking at your waste in new ways.

Throw everything but the kitchen sink into the pot, and you'll turn out a savory stock that's seasoned exactly how you like it. Keep in mind that if your beets are red or purple, they'll give a reddish tint to the stock.

Cruciferous vegetables (such as cabbage, kale, collard greens, broccoli, and Brussels sprouts) and other members of the *Brassicaceae* family (such as radishes, rutabaga, and turnips) tend to overpower other vegetables in a stock, imparting a bitter or cabbage-y flavor, so avoid them (or use them sparingly) unless you want to make a cabbage-flavored stock. Same goes for artichokes, which have a strong, distinctive aroma that doesn't play well with others.

THE ONION GROUP	THE SWEET GROUP	THE VEGETAL GROUP	THE SEASONING GROUP
Onion ends, skins, and outer layers	Carrot peels and ends	Celery ends and leaves	Unpeeled garlic
Shallot ends, skins, and outer layers	Tomato peels, cores, and ends	Asparagus peels and ends	Carrot greens
Leek ends and leaves	Bell pepper ends	Zucchini cores and ends	Parsley stems
Scallion ends	Lettuce spines and outer leaves	Parsnip cores and ends	Leftover herbs
Chive blossoms	Pea pods	Beet ends	Ginger ends
	Corn cobs	Cucumber peels	Dried mushrooms
	Fennel fronds, stems, cores, and ends	Chard stalks	Dried kelp
		Fava bean pods	Bay leaves
		Green bean ends	Peppercorns
		Mushroom stems	Leftover tomato paste
		Potato skins	

THE BASICS OF STOCK MAKING

Using the chart on page 27, choose a few ingredients from each column to create a balanced stock. I tend to go heavy on the onion and sweet groups, moderate with the vegetal group, and light on the seasoning group.

Chop up larger pieces of vegetables to increase their surface area (and enhance their ability to flavor the stock) and place them in a large pot. Cover with a few inches (7.5 to 10 cm) of water and bring to a boil.

Reduce the heat and simmer for one hour. Strain the stock (I find it easiest to pour the liquid through a colander or sieve into another pot) and let it cool to room temperature. Compost the scraps, which probably resemble a vegetable graveyard at this point.

Store the stock in pint (473 ml) or quart (946 ml) jars (depending on how you usually use it) and refrigerate or freeze. Fresh homemade stock should last about a week in the fridge and several months in the freezer.

TIPS FOR A FULLER FLAVORED VEGETABLE STOCK

Brown Your Aromatics

To give your stock an initial flavor boost, brown your aromatic vegetables (onions, carrots, and celery, or their equivalent) before adding water. Heat a wide, heavy pot over medium-high heat, drizzle in a tablespoon (15 ml) of olive oil, then add your vegetables. Throw in a few pinches of salt and cook, stirring occasionally, until the vegetables begin to brown. Pour in the water, covering the vegetables by a couple of inches (2.5 to 5 cm), then add any remaining ingredients such as garlic, herbs, mushroom soaking liquid, or leftover tomato paste.

Don't Be Shy with Mushrooms

Mushrooms are rich in glutamic acid, a natural version of the flavor enhancer monosodium glutamate (MSG), which adds umami flavor to stock. Unlike MSG, the naturally occurring glutamic acid does not contain high levels of sodium. You can add fresh mushrooms or dried mushrooms to deepen the savory notes in the stock. If using dried mushrooms, soak them in hot water first and add both the rehydrated mushrooms and soaking liquid to the pot.

Amp Up the Umami

For a richer flavored stock, add other glutamic acid-heavy ingredients such as ripe tomatoes, Parmesan cheese rinds, or dried kelp (also known as dashi kombu and frequently used in Japanese cooking).

A PRIMER ON PESTO

Thumbing through this book, you might notice there is a lot of pesto, especially pesto made from things *other* than basil. I often like to joke if it's green, it can be turned into pesto; but that's not too far off.

Pesto is derived from the Italian word *pestare*, meaning to pound, and it refers to the traditional method of preparation with a marble mortar and a wooden pestle. When we think of pesto, visions of the classic Italian sauce come to mind, but myriad versions of this "pounded sauce" exist. Some stay true to the original Genovese recipe of garlic, pine nuts, basil, Parmigiano-Reggiano, and olive oil, while others are inspired by those ingredients but have a character all their own. These days, pesto has become a generic term for any kind of sauce that's been pounded (or processed) into a thick, green condiment.

You can make pesto with many other herbs besides basil, as well as other leafy greens. You can substitute crumbly Cotija for grated Parmesan or pumpkin seeds for pine nuts. You can omit the cheese entirely or add other spices to give it more heat. And you can make the pesto as thin or as thick as you like, to use as a sauce for pastas and pizzas or a spread for sandwiches and crostini.

I always keep a jar of pesto in the fridge for its versatility; it can elevate the blandest of dishes into the most flavorful of concoctions. To perk up a pot of minestrone, stir in a few spoonfuls of pesto. If you want to add a little special something to chicken, steak, potatoes, or carrots, pesto makes an easy and elegant sauce for drizzling.

Preparing your own pesto out of the odds and ends from your pantry and produce bin is simple once you know the formula. The proportion of greens can vary by up to a cup (235 ml), give or take, as it depends on whether you use the tender leaves of vegetables or their denser stems and stalks. Stronger-flavored herbs, like sage and tarragon, should be paired with milder-tasting herbs like parsley and basil. Stems from a few different greens can be tossed in together, so save your stems throughout the week and turn them into an "anything goes" stem pesto. I often add a small handful of herbs or greens (like cilantro or spinach) to my stem pesto for a smoother texture.

HERBS OR GREENS		CHEESES	NUTS OR SEEDS
Basil leaves and flowers	Nasturtium leaves and flowers	Parmesan	Pine nuts
Parsley leaves and flowers	Dandelions	Pecorino	Walnuts
Cilantro leaves and flowers	Arugula	Grana Padano	Peanuts
Mint leaves and flowers	Spinach	Asiago	Pecans
Sage leaves and flowers	Watercress	Cotija	Almonds
Thyme leaves and flowers	Mache (corn salad)		Pistachios
Tarragon leaves and flowers	Miner's lettuce (claytonia)		Cashews
Dill fronds and flowers	Carrot greens		Macadamia nuts
Fennel fronds and flowers	Kale and/or kale stems		Hazelnuts
Rosemary flowers	Chard and/or chard stems		Sunflower seeds
Oregano flowers	Collard greens and/or collard stems		Pumpkin seeds
Marjoram flowers	Broccoli stems		Sesame seeds
Chive blossoms	Cauliflower stems		Hemp seeds
Garlic chives	Shelled peas and/or pea shoots		
Scallions	Shelled fava beans and/or fava leaves		
Ramps	Bean leaves		
Green garlic shoots			
Garlic scapes			

A FOOLPROOF FORMULA FOR PESTO

MAKES 1 CUP (235 ML)

2 cups (weight will vary) packed herbs or greens

½ cup (weight will vary) grated hard cheese

⅓ cup (weight will vary) toasted nuts or seeds

3 garlic cloves

¼ to ½ cup (60 to 120 ml) oil

Salt to taste

PESTO VINAIGRETTE

For a quick and simple salad dressing, start with a smooth, thin pesto (or add more oil as needed to make it pourable). Whisk together ¼ cup (60 ml) pesto and 2 tablespoons (28 ml) wine vinegar, taste, and add more vinegar if desired.

METHOD AND STORAGE

Pesto is typically made one of three ways. The traditional method comes from its namesake and involves pounding the ingredients together with a mortar and pestle. Purists may insist this is the only way to properly release and blend all the flavors and oils, and to a certain extent, this is true. But in a modern kitchen, a food processor or blender does the job quite well, and it's my go-to method in this book. You can also mince and combine all the ingredients with a knife or mezzaluna, adding a drizzle of olive oil in the end to hold the sauce together.

If you don't use all the pesto right away, pack it tightly into an airtight jar and pour a thin layer of olive oil on top before refrigerating. The oil will help prevent the pesto from oxidizing and turning brown (though it's still perfectly edible when it's discolored).

You can also freeze the pesto for future use, especially when you're craving a fresh burst of summer flavor in the dreary winter months.

THREE EASY WAYS TO FREEZE PESTO

1 Pour the pesto into a resealable plastic bag, flatten it out, and freeze. If you're making large batches of pesto, you can stack individual bags of one-cup (235 ml) servings on top of each other.

2 Use a small cookie scoop or a wide spoon to scoop single portions of pesto onto a parchment-lined baking sheet. Place the baking sheet in the freezer for several hours until the pesto balls are firm. Store them all in a freezeproof container and thaw only what you need each time.

3 Pour the pesto into an ice cube tray and freeze. Each mold is usually a tablespoon's (15 ml) worth, and you can pop out a pesto cube anytime you need one.

TOOL TIP

On most modern food processors, the plunger that fits inside the feeding tube has a tiny hole at the bottom, and it actually serves a purpose. Pour your oil into the plunger and let it slowly stream into the bowl while the processor is running. It saves you the work of drizzling in the oil yourself.

2

TOMATOES & PEPPERS

Of all the things that grow in my garden in the summer, I'm especially partial to my nightshades. And by nightshades, I don't mean Belladonna—or Atropa belladonna, otherwise known as the infamous deadly nightshade.

Tomatoes and peppers belong to the nightshade family as well, but are hardly the deadly nightshade most people think of when they hear the term. Their association is unfortunate, as it means many people have shied away from cooking and eating the leaves of tomatoes and peppers. Granted, these are not the kinds of greens you're likely to find at a farm stand, but for those of you lucky to have them in your yard, they're worthy of a taste.

It's difficult to describe what a tomato leaf tastes like. When I try to envision it, I always come back to the most recognizable scent of summer: sharp, slightly tangy, and almost herbal, filling the air as I brush against the thick leaves to pick a ripe, juicy tomato warmed by the sun. Tomato leaves taste, well, like a deep, earthy vine.

As for pepper leaves, they're mildly pungent, like white pepper. They're primarily used to accent a dish (as you would use white pepper), which is a good thing, since you wouldn't want to strip all the leaves off a fruiting plant. When both are used in moderation, they add a distinct depth of flavor to food that you can't find in the fruits alone.

SPICY MINTY TOMATO SAUCE

INFUSED WITH TOMATO LEAVES

MAKES 4 CUPS (946 ML)

2 pounds (900 g) tomatoes, chopped

½ cup (30 g) packed fresh parsley

⅓ cup (12 g) packed fresh mint

4 garlic cloves

¼ cup (60 ml) olive oil

1 tablespoon (6 g) ground black pepper

½ teaspoon red pepper flakes

⅓ cup (18 g) packed tomato leaves and stems

I used to make this same sauce, years ago, without the tomato leaves. Then I discovered tomato leaves were edible and imagined that earthy, heady, vine-y scent—which always makes me think of summer—in a sauce. Tomato leaves truly add another dimension to this sauce, making it richer, more fragrant, and more tomato-y. Like good wine, the aroma is subtle; fresh parsley and mint hit the nose first, followed by a lingering spice on the taste buds, and finally a smoother, deeper finish with hints of sun-kissed tomatoes still on the vine. If you don't have access to this secret ingredient, the sauce is nearly as spicy, minty, and fresh without it. Think of it as a bold take on the classic arrabbiata sauce.

In a blender, puree the tomatoes, parsley, mint, garlic, and oil until smooth. Pour the sauce into a medium saucepan and place over medium heat. Stir in the black pepper (don't be afraid, put it all in there!) and red pepper flakes. Add the tomato leaves and simmer uncovered, stirring occasionally until the sauce is thickened and infused with flavor, about 20 minutes. Remove the tomato leaves before serving the sauce.

COOK'S NOTE
Make it easier to remove the tomato leaves from the sauce by harvesting a small handful of whole sprigs and simply eyeballing the measurement for ⅓ cup of leaves and stems.

TOMATO LEAF PESTO

MAKES 1 CUP (235 ML)

2 cups (48 g) packed fresh basil

½ cup (50 g) grated Parmesan cheese

⅓ cup packed tomato leaves

¼ cup (35 g) toasted pine nuts

3 garlic cloves

¼ teaspoon kosher salt

¼ to ½ cup (60 to 120 ml) olive oil

When I think of summer bounties, I think of basil and tomatoes—the poster children for the season. Something about the sweet, savory, and ever-so-slightly peppery aroma of basil makes a fruity, subtly smoky, vine-ripened tomato sing. When you combine both of their characteristics into an otherwise traditional pesto, the result is a sauce that is unmistakably basil-scented, but with a note of warm and earthy tomato leaf.

Slather it on a thick slice of mozzarella for a different take on the classic Caprese salad.

Add the basil, Parmesan, tomato leaves, pine nuts, garlic, and salt to a food processor and pulse until crumbly, scraping down the sides of the bowl with a rubber spatula as needed. Continue pulsing and add the oil in a steady stream until well blended. Use ¼ cup (60 ml) oil for a thicker paste or up to ½ cup (120 ml) oil for a thinner sauce.

SPICY FERMENTED
SUMMER SALSA

MAKES 4 CUPS (946 ML)

1½ pounds (680 g) tomatoes, cut into small dice

½ red onion, cut into small dice

½ to 1 jalapeño pepper, minced (some jalapeños are hotter than others, so do a taste test before putting the whole thing in there)

½ serrano pepper, minced

5 garlic cloves, minced

½ cup (8 g) chopped cilantro

1½ teaspoons kosher salt

½ teaspoon ground cumin

Juice of 1 lime

Olive oil for topping

Why fermented salsa? Why not normal salsa like you've always made? I've used this same recipe for nonfermented salsa and it's fine—great, actually—but fermentation pushes it over the line to fantastic.

The same bacteria and yeasts that give sauerkraut and kimchi their distinctive flavor also give this salsa a bright and tangy note. And, those same bacteria and yeasts are what make this salsa so good for your gut. It's an easy way to get fermented foods in your diet if you're not keen on kraut; the salsa is lively on the taste buds without being too sour or too salty.

In a medium bowl, combine all the ingredients (except the oil). Pour the salsa into a quart jar (946 ml) and run a knife around the sides of the jar to release any trapped air bubbles. Add a ½-inch (1.3 cm) layer of oil on top. The oil serves two purposes: It prevents the vegetables from rising above the liquid and growing mold on the surface, and it adds a richness to the salsa once you mix it in.

Loosely seal the jar with a lid and let it stand at room temperature, out of direct sunlight, for a few days. The warmer your room temperature is, the faster your salsa will ferment. You'll begin to see bubbles in the jar as the lactic acid bacteria flourish. Taste the salsa after 3 to 4 days; if it hasn't developed a bold, tangy flavor yet, leave it out for up to 1 week. The longer you let it ferment, the more intense the flavor will become, and the longer the salsa will keep (since the bacteria are a natural preservative). Refrigerate once the salsa reaches peak flavor. The oil may congeal in the cold temperature, but it is still safe to eat. Bring the salsa to room temperature and stir in the oil before serving.

EASY DOES IT

Fermented salsa undergoes the same process of lacto-fermentation as sauerkraut (the lacto refers to lactic acid, not lactose). By letting your salsa sit out for a few days, you'll encourage all kinds of beneficial bacteria in the mix, creating a powerful probiotic that you can't get enough of (in my house, a jar never lasts more than a couple of days!). Many fermentation recipes call for the addition of whey or starter culture to infuse the food with good bacteria, but this simple recipe requires only the existing bacteria (which are already present on the skins of all your vegetables) to get started. It may take a day or two longer to ferment, but the ease makes it worth the wait.

TOMATO SOUP
TOPPED WITH OLIVE OIL–TOASTED PINE NUTS

MAKES 6 TO 8 SERVINGS

2 tablespoons (28 ml) olive oil, divided

½ pound (225 g) leeks (white stems and green leaves), sliced

4 garlic cloves, chopped

4 pounds (1.8 kg) tomatoes, chopped

3 cups (700 ml) chicken broth

⅛ teaspoon ground black pepper

1 cup (24 g) packed fresh basil

½ cup (50 g) grated Parmesan cheese

1 cup (135 g) pine nuts

2 tablespoons (8 g) minced fresh parsley

Zest of 1 lemon

I've always wondered why many tomato soup recipes call for peeling the tomatoes. The skins impart such a rich flavor on their own, it seems a waste to discard them for the sake of smoothness. Adding an immersion blender to your kitchen arsenal will transform even the toughest skins into silky goodness. You'll be blending your way into velvety soups and sauces while keeping the extra sweetness that comes from the skins.

Heat a large pot over medium-high heat. Add 1 tablespoon (15 ml) of the oil and stir in the leeks and garlic. Cook until the green leeks are wilted, about 5 minutes. Add the tomatoes, broth, and black pepper and bring the liquid to a boil.

Reduce the heat and simmer uncovered until the skins soften and slide off all the tomatoes, about 15 minutes. Stir in the basil and Parmesan. Using an immersion blender, blend the soup until smooth.

Heat a separate small saucepan over medium heat. Swirl in the remaining 1 tablespoon (15 ml) of oil, and add the pine nuts, parsley, and lemon zest. Toast them in the pan, stirring frequently until the pine nuts turn golden brown and release a rich, nutty fragrance, 8 to 10 minutes. Serve the soup with a few spoonfuls of toasted pine nuts on top.

SKILLET EGGS
POACHED IN SERRANO TOMATO SAUCE

MAKES 6 SERVINGS

2 tablespoons (28 ml) olive oil

1 yellow onion, finely chopped

4 garlic cloves, minced

2 bell peppers, finely chopped

1 serrano pepper, minced

2 tablespoons (32 g) tomato paste

2 pounds (900 g) tomatoes, finely chopped

1 teaspoon red wine vinegar

1 teaspoon kosher salt

1 teaspoon smoked paprika

1 teaspoon ground cumin

1 cup (36 g) finely chopped chard

6 eggs

¼ cup (4 g) chopped cilantro

When it comes to breakfasts, I like savory, eggy, and fast. But mostly fast if I'm cooking up breakfast for a crowd (because no one likes cold anything, unless it's cereal). This one-skillet wonder is just that—a blend of spicy, savory, and saucy with half a dozen eggs to make a filling meal for the whole table.

In Tunisia, where it originated, it's known as shakshuka, slang for "mixture" in Arabic. Typically served in a cast-iron pan, shakshuka is a traditional meal of eggs poached in a medley of tomatoes, peppers, and onions simmered together. Here, I add chard to plump up the sauce and a serrano to give it some kick. Bring the whole pan to the table, hot off the stove, and serve with a loaf of crusty bread to sop up the sauce.

Heat a large skillet over medium-high heat. Coat the surface with the oil and add the onion and garlic. Cook until the onion starts to turn translucent, 2 to 3 minutes. Add the bell peppers, serrano, and tomato paste, stir to coat, and cook until the peppers start to soften, 2 to 3 minutes. Add the tomatoes, vinegar, salt, paprika, and cumin, and bring to a rapid boil. Reduce the heat and simmer uncovered, stirring occasionally until the tomato sauce is slightly thickened, about 10 minutes. Stir in the chard.

Using a spoon, make six small wells in the sauce and crack an egg into each well. Cover and cook until the yolks are just set, 8 to 10 minutes. (If you like your yolks less runny, poach for a few additional minutes.) Remove the skillet from the heat and scatter the cilantro on top.

HEIRLOOM TOMATO GALETTE
WITH TOMATO LEAF PESTO

MAKES 2 TO 4 SERVINGS

for the dough

1¼ cups (156 g) cold all-purpose flour

½ cup (112 g) cold butter, cut into small chunks

¼ cup (60 g) cold sour cream

½ teaspoon kosher salt (optional, if using unsalted butter)

¼ cup (60 ml) cold water

I used to shy away from making galettes (basically, free-form rustic tarts) because anything that required rolling out dough seemed labor-intensive, but this galette changed my mind! The dough comes together in minutes with a food processor, and you can double the portions to freeze a second batch of dough for later. My secret ingredient—sour cream—makes the crust incredibly light and flaky. (Even if you don't make a galette, this also happens to be the perfect dough for pie crust.)

The result is a buttery, juicy, rich, and savory galette piping hot out of the oven. I like to try a variety of pestos with this recipe, as they each bring a slightly different flavor to the finished galette.

Before you make the dough, make sure the flour, butter, sour cream, and water are all very cold. (I like to chill them in the freezer for about 15 minutes before I begin the recipe.)

To make the dough, add the flour, butter, sour cream, and salt (if using) to a food processor and pulse until the butter is incorporated and the mixture is crumbly. (You should still see large flecks of butter in the dough.) Add the water in a continuous stream and pulse a few more times until the dough starts to clump together but still looks a little coarse. Do not overmix, as it will make the dough too tough. When you pinch a piece of it, the dough should just stick together but still have small cracks in it. Gather the dough with your hands and shape it into a small disk about ½ inch (1.3 cm) thick. Wrap the dough tightly in plastic and refrigerate for at least 1 hour (up to 24 hours).

When the galette is ready to be assembled, preheat the oven to 400°F (200°C or gas mark 6). Line a large rimmed baking sheet with parchment paper and set aside. Sprinkle the tomatoes with the salt and let them drain on paper towels while you prepare the dough.

Remove the dough from the refrigerator. On a floured surface, roll the dough into a 12 inch (30 mm) round about ⅛ inch (3 mm) thick. Transfer the flattened dough onto the prepared baking sheet and center it on the parchment. Spread a layer of pesto on the dough, leaving a 2-inch (5 cm)

1 2 3

4 5 6

border all around. Spread the ricotta over the pesto and then layer the shallot and tomato slices on top. Fold the bare edge of the dough over the filling, moving around the circle until the dough forms a neat but rustic crust to hold in the filling. Brush the egg evenly over the crust and drizzle a little oil over the tomatoes. Bake until the crust is puffy and golden brown, 30 to 40 minutes. The galette may seem overly moist when it comes out of the oven, but let the bubbling subside for a few minutes and it will firm up. To serve, slice the galette into four wedges.

for the filling

1 pound (455 g) tomatoes, sliced into ¼-inch (6 mm) rounds

A few pinches of kosher salt

½ cup (120 ml) Tomato Leaf Pesto (page 35) or your pesto of choice

½ cup (125 g) ricotta cheese

1 shallot, thinly sliced

1 egg, lightly beaten

Drizzle of olive oil

COOK'S NOTE
If you don't have a food processor, you can make the dough by hand. Cut the butter into the flour and salt with a pastry blender until it forms small crumbles, then stir in the sour cream. Sprinkle in a bit of water at a time as you press the dough together with your (preferably cold) hands, being careful not to overhandle it. Alternatively, you can freeze the butter and then grate it with a cheese grater to make it easier to work with.

PEAK-OF-SUMMER
ROASTED RATATOUILLE

MAKES 4 TO 6 SERVINGS

1½ pounds (680 g) tomatoes, cut into 1-inch (2.5 cm) chunks

1 pound (455 g) crookneck squash, cut into ½-inch (1.3 cm) slices

1 pound (455 g) Japanese eggplant, cut into ½-inch (1.3 cm) slices

2 bell peppers, cut into 1-inch (2.5 cm) pieces

1 yellow onion, cut lengthwise into eighths

10 garlic cloves, smashed with the flat side of a knife

¼ cup (60 ml) olive oil

1½ teaspoons kosher salt

¼ teaspoon ground black pepper

1 rosemary sprig

¼ cup (10 g) thinly sliced fresh basil

Originating in France, ratatouille is a Provençal classic that puts all of summer's abundance together into one savory stew. Though this dish is traditionally sautéed, roasting the vegetables brings out a richness and sweetness that you just don't get from the stovetop. Little more is needed than a generous glug of olive oil, a fresh sprig of rosemary, and some salt and pepper to marry the flavors while they caramelize.

You can serve the ratatouille as a side dish or make it a full meal with a loaf of crusty bread and a glass of red wine. Leftovers go great on a bed of mixed greens the next day.

Preheat the oven to 400°F (200°C or gas mark 6). If your oven cannot fit two large baking sheets side by side, place one rack in the top third of the oven and one rack in the bottom third of the oven.

As you prepare all the vegetables, cut the tomatoes first and let them drain in a colander while you break down the remaining ingredients.

In a large bowl, gently toss all the vegetables with the garlic, oil, salt, and pepper until evenly coated. Strip the leaves off the rosemary sprig and scatter them on top. Spread the vegetables across two large rimmed baking sheets in a single layer, with the tomatoes cut sides up. You want the vegetables packed in tightly, but not piled on top of each other. Roast until most of the vegetables are soft, shriveled, and slightly browned, about 45 minutes. If your baking sheets are on two separate racks, swap their positions halfway through the roasting time for even cooking.

Transfer the vegetables and all their juices to a serving bowl and toss with the basil. Serve warm or chilled.

SOUTHERN
HOT PEPPER VINEGAR

MAKES 1 CUP (235 ML)

1 cup (235 ml) white vinegar

½ teaspoon kosher salt

12 fresh chile peppers (more or less, depending on pepper and heat preference)

Southern hot pepper vinegar is the ultimate hot sauce, delivering heat *and* tang in a single punch. This is the stuff that gives Southern meals their sass. Try a drizzle over my Down-Home Collard Greens (page 72), turnip greens, fried okra, fried catfish, fried chicken, or even a Bloody Mary.

This vinegar can be as mild or as fiery as you like, and the peppers you use determine the taste: jalapeños for a little heat, habaneros for a lot of heat, and serranos, cayennes, or tabascos for anything in between. You can even mix them!

In a small saucepan over medium-high heat, combine the vinegar and salt and bring to a boil. Stir until the salt is dissolved, then turn off the heat.

Trim the stems from the peppers, and either leave the peppers whole or slice them in half lengthwise. (Slicing them will produce a hotter vinegar.) Pack a bottle with the peppers, leaving about 1 inch (2.5 cm) of headspace. Pour the hot vinegar over the peppers and infuse for at least one week in a cool, dark place. The longer the peppers infuse, the hotter your vinegar will be.

ITALIAN
HOT CHILE OIL

MAKES 1 CUP (235 ML)

12 dried chile peppers (more or less, depending on pepper and heat preference)

1 cup (235 ml) olive oil

Italian hot chile oil (known in the motherland as *olio santo*, or holy oil) is found on nearly every table in Southern Italy. If you've never used it before, you'll be amazed at how a splash of hot chile oil truly adds a finishing touch to pizzas, pastas, soups, grilled meats, and roasted vegetables.

Use your choice of dried chile peppers in the infusion—the hotter the better. Over time, you might be calling this *olio diabolica* to match its hue and bite!

Trim the stems from the peppers and slice the peppers crosswise. Save any seeds that try to escape, and place all the peppers and seeds in a bottle. Pour the oil over them and infuse for at least two weeks in a cool, dark place. The longer the peppers infuse, the hotter your oil will be. If desired, strain the oil before serving (though it's customary to leave the peppers in the bottle).

COOK'S NOTE
Make sure your dried chiles are dried throughout and not simply case hardened (whereby the outside is dry and crisp, but the inside is still moist).

Italien Hot Chile Oil (left);
Southern Hot Pepper Vinegar (right)

BACON AND PARMESAN
STUFFED AND ROASTED BABY BELLS

MAKES 6 TO 8 SERVINGS

4 bacon strips

½ yellow onion, finely chopped

24 (3-inches or 7.5 cm each) baby bell peppers

½ cup (50 g) grated Parmesan cheese

½ cup (750 g) ricotta cheese

½ cup (60 g) toasted breadcrumbs

¼ cup (15 g) chopped fresh parsley

Don't you love how these adorable peppers seem to stand at attention with their hats tipped? Good looks aside, these party-ready poppers are a pleasing blend of salty and sweet. Morsels of bacon and breadcrumbs add a bit of crunch to melted Parmesan and ricotta in every bite. These tender baby bells are a different take on the traditional jalapeño poppers and make fun finger food for a crowd. (Add a chopped jalapeño or two to the filling if you still want some heat.)

Preheat the oven to 350°F (200°C, or gas mark 4).

Heat a large skillet over high heat. Fry the bacon for about 3 minutes, then flip and fry until crisp to your liking, about 2 minutes. Drain the bacon on paper towels and reserve 1 tablespoon (15 ml) of the bacon grease in the skillet.

Reduce the heat to medium and add the onion. Cook until the onion is soft and translucent, 2 to 3 minutes. Remove from heat and let cool while you prepare the remaining ingredients.

Cut the top ½ inch (1.3 cm) off the peppers, reserving the tops, then cut the bottom ½ inch (1.3 cm) so they stand upright (don't worry if you make a hole in the bottom). Scrape out and discard the seeds, then arrange the peppers evenly on a large rimmed baking sheet.

Crumble or chop the bacon into fine bits. In a medium bowl, combine the bacon, onion, Parmesan, ricotta, breadcrumbs, and parsley. Heap the filling into each pepper with a small spoon and cover with the reserved tops. Try to press as much of the filling down as possible toward the bottom of the peppers so they don't become top heavy and tip over. Roast until the peppers begin to soften, 15 to 20 minutes.

GINGER-SPICED CHICKEN SOUP WITH WILTED PEPPER LEAVES

MAKES 6 TO 8 SERVINGS

1 tablespoon (15 ml) olive oil

1 yellow onion, chopped

2-inch (5 cm) piece ginger, minced

4 garlic cloves, minced

2 pounds (900 g) skinless bone-in chicken thighs or drumsticks

8 cups (1.9 L) water

¼ cup (60 ml) fish sauce

2 teaspoons black peppercorns

1 teaspoon kosher salt

2 chayotes, cut into 1-inch (2.5 cm) chunks

2 cups (68 g) packed pepper leaves

1 lime, sliced into wedges for serving

In Filipino cooking, this soup is called *tinola,* and the stars of the simple home-style meal are the pepper leaves and chayotes. You might turn up your nose at the thought of eating pepper leaves, but unlike their fleshy counterparts, the leaves are mellow in flavor and taste pleasantly of white pepper. They're best eaten cooked, so they impart a hint of spice that seasons the whole dish. Toss in some tender chayotes, and they'll just soak up all the peppery, gingery flavors in the broth. You can serve the soup as is, but I especially like it over a bed of steamed rice.

Heat a large pot over medium-high heat. Add the oil, onion, ginger, and garlic and cook until the onion starts to turn translucent, 2 to 3 minutes. Push the mixture aside with a long spoon, add the chicken, and brown each side for about 5 minutes. Pour in the water, fish sauce, peppercorns, and salt, and bring to a boil.

Reduce the heat and simmer uncovered until the chicken is cooked through, about 20 minutes. Skim and discard the foam that rises to the surface as the chicken cooks. Add the chayotes and continue simmering for about 15 minutes, until tender and translucent. Stir in the pepper leaves and heat through until wilted. Serve with a squeeze of lime over each bowl.

COOK'S NOTE

The leaves from any sweet or hot pepper plant (*Capsicum annuum, Capsicum frutescens,* or *Capsicum chinense*) can be used in this recipe, but if you don't grow your own, you can buy them from most Filipino markets. Another peppery green like arugula or cress can also be used in place of the pepper leaves.

Chayote is a tropical fruit from the gourd family, and I always joke that it looks like a pear missing its dentures. The flavor and texture is akin to summer squash, so if you can't find chayote in your local Latin, Asian, or farmers' markets, feel free to use zucchini or another tender squash instead.

GRILLED PEPPER, PEACH, AND PORTOBELLO STACKS

MAKES 4 SERVINGS

for the marinade

¼ cup (60 ml) olive oil

¼ cup (60 ml) balsamic vinegar

Juice of ½ lemon

2 garlic cloves, minced

1 teaspoon minced fresh thyme

1 teaspoon kosher salt

¼ teaspoon ground black pepper

for the stacks

2 bell peppers, quartered lengthwise and seeded

2 peaches, halved lengthwise and pitted

4 portobello mushrooms, stems removed

4 ciabatta buns

8 ounces (225 g) mozzarella fresca, sliced

2 cups (110 g) mixed baby greens

Even if you don't "do" the grill with all of its heavy, smoky meats, you can still grill up a healthy burger (and I'm talking size here, not just nutrition) at a barbecue, and the meat lovers in the group won't miss the ground beef at all. It's hefty—piled high with marinated portobellos, peaches, and bell peppers, and stacked between two ciabatta buns with a bed of baby greens and a slice of fresh mozzarella. It's equal parts sweet and savory, and you'll get all the satisfaction of biting into a burger with juices dripping down your chin.

A stack alone makes a plentiful meal, but pair it with the Oven-Baked Potato Parmesan Fries (page 146) and a cold beer, and you have all the makings of a swell summer weekend.

In a small bowl, combine all the marinade ingredients. Place the peppers, peaches, and portobellos in a large shallow dish, pour the marinade over them, and toss to coat. Let stand at room temperature for at least 30 minutes to marinate.

Meanwhile, preheat the grill over medium-high heat. Arrange the peppers, peaches, and portobellos on the grill grate in a single layer, cover, and grill for about 5 minutes per side. What to look for: slightly charred and softened peppers, caramelized peaches with prominent grill marks on the flesh, and juicy, tender portobellos with slightly darkened and wrinkled caps. Some pieces may need less grilling time than others, so keep an eye on the ingredients and transfer them to a large dish as they finish.

Once you've taken all the peppers, peaches, and portobellos off the grill, split the ciabatta buns and place them on the grate. Grill for 1 to 2 minutes until toasty and golden brown, then flip the buns and grill the other side for 1 to 2 minutes.

To assemble the burgers, place a small handful of baby greens on each bottom bun. Stack the portobello, mozzarella, peach, and peppers on the mound of greens, and top with the remaining bun. Serve with a toothpick to hold the stack together, if needed.

3

LEAFY GREENS

It might seem odd to group broccoli with a bunch of leafy greens, but in fact, broccoli, kale, and collards all belong to the *Brassicaceae* (mustard) family, otherwise known as cruciferous vegetables or cole crops.

Broccoli is one of those special surprises for gardeners who have never grown it, as the head emerges from a stout rosette of thick leaves that sometimes span up to 2 feet (60 cm) long. We commonly consume the flower buds of the plant (what we call a "head" of broccoli), while the leaves are oddly neglected but equally delicious, appearing at farmers' markets from time to time in the spring. (Ask your favorite farmer for them!) They look like collard greens and can often be used interchangeably, but they have a distinct broccoli flavor. (And while I focus on common green broccoli in this chapter, the leaves from any broccoli, cauliflower, Brussels sprout, or kohlrabi plant can be used the same way.)

As for chard, it's actually a close relative of the beet, cultivated for its nutrient-rich leaves. Chard goes by a number of other monikers, including Swiss chard, silverbeet, spinach beet, and perpetual spinach, and it varies in color from red to yellow to purple, sometimes with stripes or streaks.

All said, this grouping of dark leafy greens is a nutritional powerhouse filled with a rich assortment of vitamins, minerals, and phytonutrients. It's what your mama always told you to eat!

KALE STEM PESTO

**MAKES 1½ CUPS
(355 ML)**

1 cup (100 g) chopped kale stems

½ cup (30 g) packed fresh parsley with stems

½ cup (50 g) toasted walnuts

3 garlic cloves

½ teaspoon kosher salt

¼ teaspoon red pepper flakes

Zest of 1 lemon

Juice of ½ lemon

¼ to ½ cup (60 to 120 ml) olive oil

How many kale recipes have you come across in which the directions tell you to reserve the stems for "another use" or even discard them altogether? Well, friends, this is your other use. Those rigid stems and ribs we often remove are the star of the show here, and they're every bit as fresh and earthy as their leafy counterparts. Used raw, they're a great pairing for the tangy lemon in this pesto.

You're not solely limited to kale stems, either; try this with your other neglected stems, such as collards, cauliflower, or broccoli.

Add all the ingredients except the oil to a food processor and pulse until crumbly, scraping down the sides of the bowl with a rubber spatula as needed. Continue pulsing and add the oil in a slow, steady stream until well blended. Some people like their pesto super smooth, but I prefer a bit of texture, so process to your liking.

For a thick paste that you can spread onto sandwiches and pizzas, use only ¼ cup (60 ml) of oil. For a thin sauce that you can stir into pastas and soups, use a full ½ cup (120 ml).

SPRING BULGUR SALAD
WITH KALE BUDS

MAKES 4 SERVINGS

for the salad

2 cups (475 ml) water

1 cup (140 g) bulgur wheat

1 teaspoon salt

4 spring radishes with greens

½ cup (73 g) shelled peas

½ cup (20 g) kale buds

½ cup (50 g) sliced scallions

½ cup (32 g) chopped parsley

for the dressing

1 teaspoon crushed garlic

1 teaspoon ground cumin

Zest and juice of 1 lemon

½ teaspoon kosher salt

¼ cup (60 ml) olive oil

Unlike kale, which can be bitter, kale buds (also called kale raab) are sweet, especially if they've been through a frost. These delicate yellow flowers appear at the end of the plant's life before it sets seed, and they actually appear on all the plants of the *Brassicaceae* family, including broccoli, Brussels sprouts, and mustard greens. They're edible and tender, and my favorite use for them is to pretty up a salad (because while many flowers are edible, most are not that palatable).

Here, I take a traditional Lebanese tabbouleh, dress it with seasonal vegetables, and go lighter on the herbs to make it more complementary to a wider variety of dishes. It's fresh, crisp, and vibrant—everything that spring is.

In a large saucepan over medium-high heat, combine the water, bulgur, and salt, and bring to a boil. Reduce the heat, cover, and simmer for about 15 minutes until the bulgur is tender. Drain and return the bulgur to the saucepan. Meanwhile, make the dressing by combining the garlic, cumin, lemon zest, lemon juice, and salt in a small bowl. Slowly whisk in the oil until well blended. Set the dressing aside as you prepare the rest of the ingredients.

Trim the greens from the radishes. Finely chop the greens and thinly slice the radishes. Add the radish greens, radishes, peas, kale buds, scallions, and parsley to the bulgur, pour in the dressing, and toss to combine. The salad can be served immediately at room temperature, but for the best flavor, refrigerate for at least 1 hour to let the bulgur soak up the dressing and herbs.

ZUPPA TOSCANA

MAKES 6 TO 8 SERVINGS

1 pound (455 g) spicy Italian sausage, casing removed and sausage crumbled

1 yellow onion, chopped

4 garlic cloves, chopped

1 russet potato, thinly sliced

¼ teaspoon red pepper flakes

8 cups (1.9 L) chicken broth

1 bunch kale, stems removed and leaves coarsely chopped

½ cup (120 ml) heavy cream

Freshly cracked black pepper to taste

I've had this recipe (and countless variations of it) in my repertoire ever since I started cooking for myself after college. I'd had some iteration of it in a few of my favorite Italian joints and wanted to recreate the same soup at home. It was hearty, easy, and economical, and one big pot could take me through a whole weekend. For an unemployed—ahem, freelancing—graduate, those were all fine points.

A decade later, you can still find me simmering a pot on a rainy weekend or a busy weeknight when I need some comforting sustenance. This classic Tuscan soup was the start of what my family likes to call my love for all peasant food. What can I say? I like to eat as the farmers do.

Heat a large pot over medium-high heat. Add the sausage and slightly flatten the crumbled pieces with a long spoon. Fry undisturbed for about 3 minutes until the sausage starts to brown on the bottom and releases its orangey-red juices. Flip all the pieces over and fry undisturbed for 2 minutes. Stir in the onion and garlic and cook until the onion starts to turn translucent, 2 to 3 minutes. Add the potatoes, red pepper flakes, and broth, and bring to a boil.

Reduce the heat and simmer uncovered for 20 minutes until the potatoes are soft. About 10 minutes before the soup is done, stir in the kale. Gently break apart some of the potatoes with a long spoon to give the broth a bit of thickness and texture. Stir in the heavy cream and heat through. Serve with a few turns of freshly cracked black pepper on top.

CHARD STALK HUMMUS

2 cups (200 g) chopped chard stalks

2 garlic cloves

¼ cup (60 g) tahini

½ teaspoon kosher salt

Juice of 1 lemon

Swirl of olive oil

Chopped fresh parsley
for garnishing

After Kale Stem Pesto (page 57), this is my favorite recipe for using up the stems and stalks of greens rather than throwing them away. Chard stalk hummus is a traditional Lebanese dish that resembles baba ghanoush in flavor and texture. In place of chickpeas, it uses chard stalks to make a rustic dip for raw vegetables and pita. You end up with a lighter texture, but all the same flavors as traditional hummus with creamy tahini, garlic, and lemon.

Conventional Swiss chard stalks will make a classic white hummus, while red or rainbow chard stalks will make a pink or yellow hummus.

Bring a small pot of water to a boil. Add the chard stalks and boil for 5 to 10 minutes (depending on how thick they are) until the stalks are very soft.

Drain well, squeezing out any excess water, and add the stalks to a food processor, along with the garlic, tahini, salt, and lemon juice. Pulse continuously until the dip is slightly chunky and still has some bite to it, scraping down the sides of the bowl with a rubber spatula as needed. Serve with a generous swirl of oil on top and a sprinkle of chopped fresh parsley, if desired.

A DIFFERENT KIND OF "NUT" BUTTER

Tahini is a paste made from ground sesame seeds. You can usually find it in a well-stocked supermarket or specialty food market with other nut butters, or in any Middle Eastern market. If you can't source tahini, try substituting natural peanut butter or another nut butter. While the hummus will still be delicious, it'll take on a slightly different flavor.

CHARD AND CHEDDAR FRITTATA

MAKES 4 TO 6 SERVINGS

1 tablespoon (15 ml) olive oil

½ red onion, thinly sliced

4 garlic cloves, minced

½ pound (225 g) new potatoes, thinly sliced

½ pound (225 g) chard, leaves coarsely chopped and stems sliced ½-inch (1.3 cm) thick on the diagonal

1 teaspoon kosher salt, divided

¼ cup (60 ml) water

8 eggs

½ teaspoon ground cayenne pepper

¼ teaspoon ground black pepper

1 cup (115 g) shredded sharp cheddar cheese, divided

A frittata is basically an Italian version of an omelette, and it's one of those kitchen pantry meals that I love to make when I need to use up a lot of odds and ends. (Soon-to-spoil vegetables, herbs, and even last night's dinner are never safe from my concoctions.) But even at its most basic—with silky Swiss chard, sweet red onion, tender potatoes, and pungent cheddar bubbling away in the oven—this meal makes breakfast an event to look forward to, and without a full production.

If you have any leftovers, turn them into frittata sandwiches (like the Italians do) for lunch the next day. They're just as good cold as they are warm.

Preheat the oven to 375°F (190°C, or gas mark 5).

Heat a large, nonstick ovenproof skillet (preferably cast iron) over medium-high heat. Coat the bottom of the pan with the oil, then add the onion and garlic and cook until fragrant, 2 to 3 minutes. Add the potatoes in a single layer, scatter the chard and ½ teaspoon salt over them, then pour in the water. Cover and cook until tender, about 5 minutes.

While the vegetables are cooking, whisk the eggs with the remaining ½ teaspoon of salt, cayenne pepper, black pepper, and ½ cup (58 g) of cheese in a medium bowl. Uncover the skillet, stir the vegetables to combine, and let any remaining water cook off.

Spread the vegetables across the skillet and pour the eggs evenly over them. As the edges of the eggs start to set, lift them with a spatula and tilt the pan slightly to let the raw eggs run off to the bottom. Cook until the eggs set on the bottom and start to set on top, about 5 minutes. Top with the remaining ½ cup (58 g) of cheese, then transfer the skillet to the oven and bake until the cheese is melted, the eggs are puffy, and the center feels firm and springy, 8 to 10 minutes. Serve the frittata right in the skillet, or invert it onto a serving platter and slice into quarters or eighths.

FRITTATA FILLINGS

The beauty of a frittata is how flexible it is. If you don't have a red onion, use green onions or leeks. To lighten up the meal, omit the potatoes and add more leafy greens. Experiment with different cheeses or mix a few together. Some of my other favorite fillings are asparagus, cauliflower, broccoli, bell peppers, mushrooms, and even leftover cooked rice or pasta.

SWISS CHARD SUMMER ROLL
WITH CITRUS SOY-MARINATED TOFU

MAKES 4 SERVINGS

for the marinated tofu

⅓ cup (80 ml) soy sauce

1-inch (2.5 cm) piece ginger, minced

Zest and juice of 1 lemon

1 block (14 ounces, or 390 g)
extra-firm tofu, sliced ½ inch
(1.3 cm) thick crosswise

for the dipping sauce

½ cup (125 g) hoisin sauce

¼ cup (60 ml) soy sauce

1½ tablespoons (20 g) sugar

1½ tablespoons (25 ml) sesame oil

2 teaspoons sriracha

Juice of 1 lemon

**for the wraps
and fillings**

16 to 18 sheets rice paper
(22 centimeter [about 8½ inches]
diameter, see sidebar)

1 bunch chard, stems removed
and leaves thinly sliced

1 bunch basil, torn into bite-size
pieces

2 avocados, sliced

2 carrots, cut into 2-inch (5 cm)
matchsticks

1 cucumber, cut into 2-inch (5 cm)
matchsticks

This cookbook wouldn't be complete without a nod to my Vietnamese roots and the salad-centric cuisine I grew up on. Vietnamese food is known for its veggie-heavy main courses and intensely aromatic herbs, and the freshness of all the meals is what makes them so light and inviting.

Summer rolls (traditionally known as *gỏi cuốn*) are an exceptional example of putting vegetables front and center of the meal. Stuffed into a slightly tangy and translucent rice paper wrap, they're like a portable salad, poised for dipping. The fillings are only as limited as your imagination allows, and the recipe here is a fusion of my upbringing. You can also try any combination of cabbage, lettuces, sprouts, peppers, jicama, summer squash, scallions, cilantro, and mint—be creative! For a touch of authenticity, add cooked shrimp and rice vermicelli noodles.

To make the marinade, combine the soy sauce, ginger, lemon zest, and lemon juice in a resealable plastic bag. Add the tofu to the bag and gently toss, making sure all the slices are coated. Let stand at room temperature while you prepare the remaining ingredients, at least 15 minutes.

To make the dipping sauce, combine all the sauce ingredients in a large bowl. Divide the sauce among smaller bowls for each guest and set aside.

Remove the tofu from the bag and discard the marinade. In a large, dry skillet over medium-high heat, fry the tofu, browning each side for about 3 minutes until a toasty crust forms. Transfer the tofu to a cutting board, cut into thinner slices lengthwise, and set aside with the other fillings.

Fill a large bowl with warm water. Using both hands, dip the rice paper in the water for a few seconds until the whole sheet is moistened. Carefully lay the rice paper flat on your work surface, smoothing out any folds, and wait a minute for the sheet to become soft and pliable.

Layer the marinated tofu, chard, basil, avocado, carrots, and cucumbers in the bottom third of the rice paper. Tightly

fold the bottom edge over the fillings and roll the paper away from you. Fold the sides in about halfway up and continue rolling to the end. Repeat with the remaining rice paper and fillings, and serve the rolls with the dipping sauce. The rice paper will start to stiffen if left out for too long, so it's best to wrap as you eat.

RICE PAPER PRIMER

Vietnamese rice paper (called *bánh tráng*) can be found in Vietnamese and Chinese markets, but these days some well-stocked supermarkets and health food grocers also carry them in their ethnic or international aisles. Rice paper sheets are made with rice, or a mix of rice and tapioca, or tapioca only (usually labeled *bánh tráng dẻo* or "tapioca sheets" on the package). The rice and rice-and-tapioca papers are a little thicker and easier to handle, while the tapioca paper is thinner, more translucent, and requires a deft hand to wrap. If you've never made your own summer rolls before, I suggest starting with an all-rice paper. I like the Three Ladies Brand in the 22 centimeter size (about 8½ inches, the same size as a regular tortilla; larger papers feel too unwieldy to wrap).

GRUYÈRE GRILLED CHEESE
WITH CHARD AND CARAMELIZED ONION

MAKES 4 SERVINGS

2 tablespoons (28 ml) olive oil, divided

½ teaspoon sugar

1 yellow onion, thinly sliced

4 cremini mushrooms, finely chopped

8 to 10 chard leaves, stems removed and leaves finely chopped

½ teaspoon kosher salt

3 tablespoons (42 g) softened butter

8 slices grainy bread

2 tablespoons (30 g) Dijon mustard

2 cups (240 g) grated Gruyère cheese

Grilled cheese has never felt like something I could cook for a crowd. By the time the last sandwiches are done, the first ones are cold or already eaten. It's a good made-to-order meal but not really conducive to a meal shared with friends. Until now!

This method for making grilled cheese eschews an actual grill in favor of the oven, which toasts the bread to grill-like perfection (sans grill marks). Creamy Gruyère and silky chard—which seems to soak up all the sweetness of the caramelized onion—turn a simple grilled cheese into an artisanal meal. You can even double this recipe to feed the whole crew, and no one gets stuck with a cold sandwich.

Place two large rimmed baking sheets inside the oven and preheat to 425°F (220°C, or gas mark 7). If your baking sheets cannot fit side by side, place one sheet on the center rack and another sheet on whichever rack remains.

In a large skillet over medium-high heat, melt 1 tablespoon (15 ml) of oil and stir in the sugar until it dissolves. Add the onion and cook, stirring frequently, until tender and translucent, about 10 minutes. Scrape up any browned bits in the skillet with a spatula and mix them back in with the onion.

Push the onions aside and add the remaining 1 tablespoon (15 ml) of oil. Add the mushrooms, chard, and salt. Cook until the vegetables are tender and wilted, about 5 minutes. Keep stirring them around to let any excess liquid cook off.

To assemble the sandwiches, thoroughly butter all the bread on one side. Turn half the slices over, buttered-side down, and spread the mustard over them. Layer equal amounts of the cheese, onion, mushrooms, and chard on each slice, then top with the remaining slices of bread, buttered-sides up.

Transfer the sandwiches to the hot baking sheet in the center of the oven and place the other hot baking sheet on top, pressing down slightly. Bake for 6 to 8 minutes until the bread is toasted on the outside and the cheese is melted on the inside.

STUFFED COLLARD GREENS

MAKES 4 TO 6 SERVINGS

for the sauce

1 tablespoon (15 ml) olive oil

1 shallot, minced

4 garlic cloves, minced

1 can (28 ounces, or 780 g) of crushed tomatoes

1 thyme sprig

¼ teaspoon red pepper flakes

¼ teaspoon kosher salt

⅛ teaspoon ground black pepper

for the filling

16 collard leaves (8 to 10 inches [20 to 25.5 cm] long), stems removed, plus more for lining

1 tablespoon (15 ml) olive oil

1 yellow onion, finely chopped

4 garlic cloves, minced

1 tablespoon (16 g) tomato paste

½ pound (225 g) ground turkey

2 carrots, finely chopped

¼ teaspoon ground black pepper

1 bay leaf

1½ cups (293 g) uncooked white rice, rinsed

3 cups (700 ml) chicken broth, divided

Sour cream for serving

Collard greens are great for stuffing. The leaves are sturdy and wide, they don't shrink after blanching, and they don't break apart after filling. They also add a vibrant green color and mildly green flavor to a pan of these *golubtsy*-inspired rolls.

"What are *golubtsy?*" you might ask. *Golubtsy* are stuffed cabbage rolls, classic favorites in every Russian family's repertoire of home-cooked meals. They're common throughout Eastern Europe and have as many variations as there are families making them. I've borrowed a bit from the traditional savory filling of meat, carrots, and onions, but top them with a zesty tomato sauce and, of course, wrap them in large, billowy leaves of collards. Don't omit the sour cream; the tangy creaminess makes this dish.

To make the sauce, heat a medium saucepan over medium-high heat. Add the oil, shallot, and garlic and cook until fragrant, 2 to 3 minutes. Stir in the remaining sauce ingredients and bring to a boil. Reduce the heat and simmer uncovered, stirring occasionally until the sauce is infused with the herbs and spices, about 20 minutes. Turn off the heat and set aside.

Meanwhile, bring a wide, shallow pot of water to a boil. Add a small stack of collard greens and blanch for about 1 minute until the leaves are pliable. Transfer to a colander to drain, then blanch the next set of greens and repeat until all of them are softened.

Empty the pot of water and return it to the stove over medium-high heat to make the filling. Add the oil, onion, and garlic, and cook until tender and translucent, 2 to 3 minutes. Stir in the tomato paste to coat. Add the turkey, carrots, pepper, and bay leaf, and cook until the turkey is no longer pink, about 5 minutes. Add the rice and stir to coat. Pour in 2 cups (475 ml) of broth and bring to a boil.

Reduce the heat and simmer uncovered, stirring occasionally until the liquid is absorbed, 12 to 15 minutes. Don't worry about cooking the rice completely; it will finish cooking in the oven. Remove the pot from heat and set aside.

Remove and discard the bay leaf.

Preheat the oven to 350°F (180°C, or gas mark 4). Line the bottom of a 9-inch by 13-inch (7.5 cm) (23 by 33 cm) baking dish with a layer of collard greens (to prevent sticking). If you don't have enough collard greens, you can use other leaves to line the dish, like cabbage or chard. (Scraps and leftovers are great for this.)

To assemble the collard rolls, place a blanched collard leaf on your work surface and scoop about ¼ cup (60 ml) of filling (more or less, depending on the size of your leaves) onto the bottom third of the leaf. Fold the bottom edge up over the filling, roll the leaf away from you, then fold the sides in and continue rolling to the end. Place the collard roll in the prepared baking dish and repeat with the remaining collard leaves.

Once all the collard rolls are stuffed and arranged in a single layer in the baking dish, pour the tomato sauce evenly over them. Add the remaining 1 cup (235 ml) of broth to the dish to steam them as they cook, and bake for 45 minutes. Serve warm with a spoonful of sour cream on top.

COOK'S NOTE
If your collard greens have thick center ribs, shave them down with a paring knife (similar to shaving the ribs on broccoli greens on page 81) to make the leaves more pliable for rolling.

DOWN-HOME
COLLARD GREENS

MAKES 6 TO 8 SERVINGS

2 pounds (900 g) collard greens

1 tablespoon (15 ml) olive oil

1 yellow onion, chopped

4 garlic cloves, chopped

1½ pounds (680 g) smoked ham hocks or smoked ham shanks

8 cups (1.9 L) water

⅓ cup (80 ml) cider vinegar

1 teaspoon red pepper flakes

¼ teaspoon ground black pepper

Kosher salt to taste

Southern Hot Pepper Vinegar (page 46) for serving

My love for collard greens started on a cross-country road trip after college when I experienced my first honest-to-goodness, stick-to-your-ribs, home-cooked-in-a-restaurant meal in the South. The bowls of bitter greens were tender and juicy with plump and smoky chunks of ham clinging to the leaves in every bite. From that point on, I was spoiled by my first mess of greens. It was what soul food should be: simple, nourishing, and meant to be shared.

Slow-simmered collard greens and savory smoked ham go hand in hand, and buying good-quality ham will make a big difference in how your collards taste. If you can, find a butcher who will chop up the ham and give you a piece of the hock (which is bonier) and a piece of the shank (which is meatier). I love this dish as a side, but you can also eat it as a hearty, head cold-easing, antioxidant-packed stew, soaked up with a chunk of cornbread.

To prepare the collard greens, trim the thick, rigid stems from the leaves. Slice the stems ½ inch (1.3 cm) thick on the diagonal and set aside. Stack a few leaves on top of each other and roll them lengthwise into a tight cigar. Slice the leaves crosswise about 1 inch (2.5 cm) thick. Repeat with the remaining leaves until you have a mountain of collard ribbons. Don't be intimidated by how huge it looks; the collards will cook down to at least half their volume.

In a wide, heavy pot over medium-high heat, add the oil, onion, and garlic and cook until the onion starts to turn translucent, 2 to 3 minutes. Add the ham hocks, water, vinegar, red pepper flakes, and black pepper, stir to combine, and increase the heat to high. Add a couple of large handfuls of the collard leaves and stems and push them down into the liquid with a long spoon. As they start to soften, add a couple more handfuls. By the time the pot reaches a continuous boil, all the collards should be submerged in liquid.

Reduce the heat, cover, and simmer, stirring occasionally until the leaves are wilted, the stems are tender, and the cooking liquid tastes smoky and earthy, about 1 hour. The longer you simmer the pot, the silkier the collards will be. Don't add any salt to this recipe until you taste it, since the salt level greatly depends on how salty the ham hocks are and how long you cook them.

Remove the ham hocks from the pot and pull the meat off the bones. Finely chop the meat, then stir it back into the pot and heat through. Serve with a drizzle of Southern Hot Pepper Vinegar (page 46) on top or substitute any tangy red hot sauce of your choice.

LIKKER UP!

If it looks like you're making a soup, you are! (Or you will be.) Don't throw out the cooking liquid, known affectionately in the South as potlikker, once you've eaten all the greens. The savory, smoky broth makes an excellent base for soup. In fact, you'll find a recipe for Potlikker Noodles with Collard Greens (page 75), a twist on the traditional potlikker that uses the leftover broth.

POTLIKKER NOODLES
WITH COLLARD GREENS

MAKES 4 SERVINGS

12 ounces (340 g) dried egg noodles

4 bacon strips, finely chopped

½ red onion, thinly sliced

2 garlic cloves, minced

8 collard leaves, ribs and stems removed, leaves coarsely chopped

4 cremini mushrooms, thinly sliced

3 to 4 cups (700 to 946 ml) potlikker from Down-Home Collard Greens (page 72)

Lemon wedges for serving

Grated Parmesan cheese for serving

Potlikker is a Southern term for the liquid produced from any mess of greens, be it turnip, mustard, or our beloved collards here. The liquid is rich with all the vitamins that leach out of the greens during the cooking process. Southern folklore even tells tales of doctors prescribing potlikker for any number of ailments, including colic and anemia.

While the healing powers of potlikker have yet to be validated, there's no denying that the slightly bitter broth is high in iron, calcium, and vitamins A, C, and K. If you're feeling a little under the weather, save the potlikker and sip it as a soup. It's comforting regardless of what ails you.

Bring a large pot of salted water to a boil and cook the egg noodles according to the package directions. Drain and set aside.

Heat a large pot over medium-high heat. Add the bacon and fry for about 2 minutes until the fat is rendered. Add the onion and garlic and stir to coat in bacon grease. Cook until the onion starts to turn translucent, 2 to 3 minutes. Add the collard greens and mushrooms and cook until tender, 6 to 8 minutes. Pour in the potlikker and heat through.

To serve, divide the noodles among four bowls and top with equal amounts of vegetables and potlikker. Squeeze a wedge of lemon over each bowl and sprinkle some Parmesan on top.

ROASTED ROMANESCO AND BROCCOLINI SALAD
WITH WILTED ARUGULA

MAKES 4 SERVINGS

½ cup (120 ml) olive oil

4 garlic cloves, crushed

1 tablespoon (2 g) chopped fresh rosemary

½ pound (225 g) Romanesco broccoli florets

½ pound (225 g) broccolini, cut into bite-size pieces

1 leek (white stem only), thinly sliced

A few pinches of kosher salt

A few grinds of black pepper

Zest and juice of ½ lemon

2 cups (40 g) packed arugula, coarsely chopped

1 tablespoon (15 ml) balsamic vinegar

½ cup (60 g) crumbled blue cheese

A broccoli salad is likely not the first thing that comes to mind when you think of a fresh, flavorful, senses-tingling salad. But a *roasted* broccoli salad, on the other hand, has all these traits and then some.

Infused with garlic-and-rosemary oil and tossed with wilted leeks and arugula, it's the kind of salad that stands out on its own, not as just another starter or side dish. Crumbled blue cheese adds a touch of saltiness and creaminess to this tangy and savory salad, which is best served warm as it melts on the palate.

Preheat the oven to 425°F (220°C, or gas mark 7).

In a small saucepan over medium heat, bring the oil to a gentle simmer and infuse the garlic and rosemary for 2 to 3 minutes, stirring once. Remove from heat.

On a large rimmed baking sheet, toss the broccoli, broccolini, and leek with ¼ cup (60 ml) of the infused oil. Scatter the salt and pepper on top and arrange the vegetables in a single layer across the baking sheet. Roast until the broccoli stalks are tender and the tips of the florets are slightly blackened, 15 to 20 minutes. Transfer the vegetables to a large serving bowl and toss with the lemon zest and arugula; the leaves will naturally wilt from the heat.

In a small bowl, combine the vinegar and lemon juice. Slowly whisk in the remaining ¼ cup (60 ml) of infused oil until well blended. Toss with the vegetables and serve with some blue cheese crumbled on top.

COOK'S NOTE
Any variety of broccoli or cauliflower works well in this recipe, so feel free to experiment with common green broccoli, sprouting broccoli, baby broccoli, and the various colors and cultivars of cauliflower.

SRIRACHA-ROASTED
BROCCOLI

MAKES 4 TO 6 SERVINGS

3 tablespoons (45 ml) olive oil

2 tablespoons (28 ml) sriracha

1 tablespoon (15 ml) soy sauce

2 teaspoons sesame oil

1 teaspoon crushed garlic

1 pound (455 g) broccoli, cut into bite-size pieces

½ lemon, cut into wedges

Broccoli in the raw is a bit tree-like, both in looks and in taste, and it takes a little seasoning to give it some character. In its virgin form, broccoli is a blank canvas, lending itself to being dressed, dipped, and marinated every which way. But I like to roast it under high heat, which brings out its inherent sweetness—a sweetness that balances the umami from this sriracha-spiced sauce.

I also like to roast broccoli on the most battered and blackened pan in my kitchen, as the broccoli seems to caramelize better, producing beautiful bits of brown that are full of flavor. Don't toss the thick stem—if it seems too fibrous for your liking, peel the skin before you roast it.

Preheat the oven to 425°F (220°C, or gas mark 7).

In a large bowl, stir together the olive oil, sriracha, soy sauce, sesame oil, and garlic. Add the broccoli and toss to thoroughly coat.

Spread the broccoli across a large rimmed baking sheet in a single layer and roast for 10 minutes. Stir and shake up the broccoli a bit and roast for 10 minutes more until the stems are tender and the tips of the florets are slightly blackened. Serve with a couple of squeezes of lemon on top, and if you really want to go to spice town, serve some more sriracha on the side for dipping.

BROCCOLI GREEN AND
BAKED FALAFEL WRAP

MAKES 4 TO 6 SERVINGS

for the falafels

2 cans (15 ounces, or 425 g each) of chickpeas, drained and rinsed

⅓ cup (55 g) finely chopped red onion

⅓ cup (20 g) packed fresh parsley with stems

⅓ cup (5 g) packed cilantro with stems

4 garlic cloves

1 egg, beaten

2 tablespoons (28 ml) olive oil

2 tablespoons (16 g) all-purpose flour

1½ teaspoons ground cumin

1 teaspoon kosher salt

½ teaspoon ground cayenne pepper

1 teaspoon baking soda

ingredients continue

Falafels are one of my favorite fast foods. I know it doesn't seem fast when you look at all the ingredients here, but you'll be surprised at how quickly it all comes together when most of it is just blitzed in a food processor.

Falafels are typically deep-fried balls of goodness, but these are baked to make a more healthful version. In the oven, they turn crisp on the outside, tender on the inside, and lighter than traditional falafels. Stuff them into the sturdy, billowy leaves of a broccoli plant, which are perfect for wrapping. Broccoli greens, which include the large outer leaves of the plant and not just the ones that encircle the head, taste like a milder version of broccoli and give this green wrap a fresh, earthy flavor.

Preheat the oven to 375°F (190°C, or gas mark 5). Line a large rimmed baking sheet with parchment paper and set aside.

To make the falafels, add all the ingredients (except the baking soda) to a food processor and pulse a few times until the mixture is well combined but still crumbly, scraping down the sides of the bowl with a rubber spatula as needed. Pulse the mixture just enough to reach a coarse, paste-like texture, but not so much that it turns into hummus. Sprinkle the baking soda evenly over the mixture, then stir it in with a fork until incorporated.

Using your hands, roll the mixture into 20 round falafels, about 1½ inches (1.3 cm) in diameter (the size of a golf ball), and place them on the prepared baking sheet. Bake for 20 to 25 minutes until the falafels are browned and crispy, turning them over halfway through for even browning on all sides.

While the falafels are baking, make the salsa by combining all the ingredients in a medium bowl and set aside. Then, make the tahini sauce by combining all the ingredients in a small bowl and set that aside.

recipe continues

1   2   3

for the salsa

2 cups (300 g) halved cherry tomatoes

1 cup (135 g) diced cucumber

⅓ cup (55 g) thinly sliced red onion

¼ cup (15 g) chopped fresh parsley

2 tablespoons (28 ml) olive oil

½ tablespoon minced garlic

¼ teaspoon kosher salt

⅛ teaspoon ground black pepper

for the tahini sauce

½ cup (120 g) tahini

½ cup (120 ml) water

3 garlic cloves, crushed

¼ teaspoon kosher salt

Juice of 1 lemon

4 to 6 broccoli greens, stems removed

continued

If your broccoli greens have thick center ribs, use a paring knife to carefully shave down the ribs on the undersides of the leaves. This will make the leaves more pliable and much easier to roll. To assemble the wrap, layer the falafels and salsa down the middle of a broccoli leaf, then drizzle a generous amount of tahini sauce on top. Fold the bottom third of the leaf up, fold each side in, then serve. The falafels can also be made ahead of time, refrigerated, and then reheated or served cold.

COOK'S NOTE
In a falafel, the difference in flavor between cooked (canned) and uncooked (soaked) chickpeas is debatable. The issue really comes down to texture. Dried chickpeas that are soaked overnight, drained well, and left uncooked hold together better when formed into balls or patties. If you prefer to use dried chickpeas, soak about 1⅓ cups (267 g) to yield the equivalent of two 15-ounce (425 g) cans. Pulse the uncooked chickpeas in a food processor with the other falafel ingredients as instructed, but omit the egg and flour (which are simply binding agents for canned, cooked chickpeas).

If you don't grow or can't source broccoli greens for wrapping, collard greens are an easy substitute. But if you have cauliflower, cabbage, or Brussels sprouts growing in your garden, these green wraps make excellent use of the underappreciated leaves. You can safely harvest a couple of large outer leaves from the plant each week without affecting the growth of the head.

4

PEAS & BEANS

Botanically, peas and beans belong to the same family of plants (*Fabaceae*) and are known as legumes. They're among my favorite plants in the garden because they're both a fruit (the pods and seeds) and a vegetable (the leaves), meaning every part of the plant is edible.

Pea vines are especially prolific as they climb up stakes several feet high. The leaves taste just like the pods, but the most tender parts are the tips of the vines, called pea shoots. Often they'll come with a few tendrils or flowers or perhaps a baby pod. The new growth is fresh and light, tasting like young peas.

Bean leaves, on the other hand, are deep and earthy. They include the leaves of common snap beans as well as fava beans, runner beans, and winged beans. Farmers don't often harvest the leaves, as the plants are valued for their pods, but if you grow your own at home (especially fava beans, which have much more foliage than other varieties), you can use the supple leaves on mature plants or the seedlings that you thin out.

When cooking with peas and beans, think beyond the green: Golden snow peas, purple-speckled snap beans, and crimson-colored pods all add whimsy to a plate.

PEA SHOOT SALAD
WITH RADISH AND CARROT

MAKES 4 SERVINGS

for the salad

2 spring radishes with greens

1 carrot with greens

4 cups (340 g) packed pea shoots, cut into 2-inch (5 cm) lengths

for the dressing

2 tablespoons (28 ml) white wine vinegar

1 teaspoon stone-ground Dijon mustard

1 teaspoon honey

A few grinds of black pepper

¼ cup (60 ml) olive oil

Shaved Parmesan cheese for serving

Tender-crisp radishes, slender young carrots, and delicate pea shoots all hold promise of a bountiful season ahead. The flavor is light and fresh, with a subtle bite from the peppery radishes.

For a salad that's almost too pretty to eat, choose pea shoots with plentiful curlicues and garnish the plates with pea flowers.

Trim the greens from the radishes. Coarsely chop the radish greens and toss them into a large serving bowl.

With a mandoline or a steady hand, thinly slice the radishes and add them to the bowl. Trim the greens from the carrot and remove the tender leaves from the rigid main stems; discard the stems. Finely chop the carrot leaves and toss into the serving bowl. Julienne the carrot (or cut into 2-inch [5 cm] matchsticks) and add the strips to the bowl. Add the pea shoots and toss to combine.

To make the dressing, combine the vinegar, mustard, honey, and pepper in a small bowl. Slowly whisk in the oil until well blended. Dribble the dressing over the salad and lightly toss. Finish with a small handful of shaved Parmesan on top and serve.

STIR-FRIED SNOW PEAS AND PEA SHOOTS
WITH SESAME GINGER VINAIGRETTE

MAKES 4 SERVINGS

for the vinaigrette

2 tablespoons (28 ml) rice vinegar

2 tablespoons (28 ml) sesame oil

2 teaspoons soy sauce

2 teaspoons grated ginger

for the stir-fry

1 tablespoon (15 ml) sunflower oil

¼ red onion, minced

2 garlic cloves, minced

2 heaping (126 g) cups snow peas

6 cups (510 g) packed pea shoots, cut into 4-inch (10 cm) lengths

Peas are sometimes grown solely for their shoots: the sweet, tender young leaves and tendrils that taste faintly of peas but can be harvested in half the time. The succulent shoots have a slightly nutty undertone and are featured prominently in Southeast Asian cooking, where they're a staple leaf vegetable. You can always find them in abundance at an Asian market, bagged together in a mess of tangles, but if a bundle turns up at the farmers' market, toss them into this simple stir-fry and serve over noodles or rice.

To make the vinaigrette, combine all of its ingredients in a small bowl and set aside.

Heat a large skillet over medium-high heat. Coat the bottom with a thin layer of the sunflower oil and add the onion and garlic, giving a quick stir until fragrant, about 30 seconds. Add the snow peas and stir-fry for 1 minute. Add the pea shoots and stir-fry for 2 minutes, shaking them up frequently for even cooking. Pour in the vinaigrette, stir well to coat, and heat through. The pea shoots are done when they've wilted to about a third of their original volume.

FRESH PEA SOUP

MAKES 6 TO 8 SERVINGS

4 tablespoons (55 g) butter

1 yellow onion, cut into small dice

4 garlic cloves, minced

6 cups (870 g) shelled peas

2 carrots, cut into small dice

1 tarragon sprig

½ teaspoon kosher salt

½ teaspoon ground black pepper

6 cups (1.4 L) water

Juice of 1 lemon

Crème fraîche for serving

Lemon zest for serving

Pea soup tends to lean toward the savory side and take on whatever flavor it's simmered with (usually ham bones or chicken stock), but this version puts that clean, fresh pea flavor front and center. The secret to making pea soup with fresh peas is using plain old water instead of stock. You don't want any other flavors to overpower the subtle sweetness of the peas. Tarragon, with its licorice-like warmth, helps bring out the sweetness as well.

In a large pot over medium heat, melt the butter. Add the onion and garlic and cook until the onion starts to turn translucent, 2 to 3 minutes. Add the peas, carrots, tarragon, salt, pepper, and water and bring to a boil. Reduce the heat and simmer uncovered until the vegetables are tender, about 20 minutes. Discard the tarragon. Ladle about half the soup into a large bowl and set aside. Using an immersion blender, puree the remaining soup in the pot until smooth.

Pour the reserved (and still chunky) soup back into the pot and stir to combine. The soup should have a soft texture that you'll get in every spoonful. Add the lemon juice and heat through. Serve each bowl with a dollop of crème fraîche and a dusting of lemon zest on top.

COOK'S NOTE

A pound (455 g) of English pea pods yields about a cup (145 g) of shelled peas. If you don't have the patience to shell them or don't have enough pods for the recipe, substitute frozen peas for all or part of the amount needed. Frozen peas tend to lack the sweetness of fresh peas, so you can wake them up with a hefty pinch of sugar in the pot. If you don't have crème fraîche on hand, substitute sour cream; the slight tang truly kicks up the flavor.

THREE BEAN SUMMER SALAD

MAKES 8 SERVINGS

for the salad

2½ pounds (1.1 kg) mixed young snap beans, cut into 2-inch (5 cm) lengths

2 cups (308 g) sweet corn kernels

2 celery ribs with leaves, thinly sliced on the diagonal

½ yellow onion, thinly sliced

½ serrano pepper, minced

for the dressing

½ cup (120 ml) cider vinegar

2 tablespoons (26 g) sugar

1 teaspoon Dijon mustard

¼ cup (60 ml) olive oil

Sometimes I feel that beans are underappreciated in the kitchen—relegated to frumpy casseroles, stuck with a reputation of being soggy and overcooked, teased as the *butt* of many jokes. But fresh beans are a lovely summer bounty, especially if you can get your hands on a medley of beautifully colored snap beans, wax beans, and French beans.

Grab three varieties (and colors) of young, slender beans with a bit of snap, whip up this salad in a matter of minutes, and be the talk of the neighborhood potluck. It travels well, feeds a whole crowd of hungry friends, and makes a light and refreshing side to all those heavy and smoky meats you find on a summer grill.

In a large serving bowl, combine all the ingredients for the salad.

To make the dressing, combine the vinegar, sugar, and mustard in a small bowl, then slowly whisk in the oil until well blended. Pour the dressing over the vegetables and toss to coat. Cover and refrigerate for a few hours to let the flavors intermingle. Serve chilled.

COOK'S NOTE
This is a great make-ahead dish, as the flavors develop more deeply the next day without turning soggy. You can use fresh or canned corn in this recipe, but if you have the time and inclination, grilled corn (shaved off the cob) would taste superb in this salad.

PAN-CHARRED BEANS
WITH BEAN LEAF PESTO

MAKES 4 SERVINGS

for the pesto

2 cups (60 g) packed bean leaves

3 tablespoons (23 g) pistachios

2 garlic cloves

½ teaspoon kosher salt

Juice of ½ lemon

6 tablespoons (90 ml) olive oil

for the beans

2 tablespoons (28 ml) sunflower oil

1 pound (455 g) French beans
(haricots verts)

Everyone eats green beans. But who eats bean greens? Well, you do now! It might seem strange to pick the leaves off plants you're normally prizing for their pods, but bean leaves are in fact edible and taste like a milder version of their podded counterparts. When pureed into a pesto, bean leaves add a deeper, greener, earthier element to these slender French beans. The bean-on-bean flavor is rich and savory, with a warm charred aroma.

If picking leaves off your own plants, spread the harvest across the whole crop so you don't affect the growth of the pods.

To make the pesto, add all of its ingredients (except the oil) to a food processor and pulse until crumbly, scraping down the sides of the bowl as needed. Continue pulsing and add the oil in a steady stream until well blended. Set aside.

Heat a wok or large sauté pan over high heat. Swirl in the oil and spread the beans across the pan in a single layer. Fry undisturbed until the beans are lightly charred on the bottom, about 2 minutes. Shake them up and cook until the beans are tender, 6 to 8 minutes, stirring occasionally. Transfer the beans to a serving dish and top with the pesto.

COOK'S NOTE

The bean leaves used in the pesto can be harvested from any common bean plant (*Phaseolus vulgaris*), as well as fava bean, lima bean, runner bean, winged bean, and yardlong bean. Young leaves are preferred for raw preparations, as they're more tender. If you don't have any bean leaves, substitute parsley or another mild green, like pea shoots or spinach.

YARDLONG BEAN CURRY
WITH WILTED SPINACH

MAKES 6 SERVINGS

1 tablespoon (15 ml) olive oil

1 yellow onion, chopped

2 lemongrass stalks, cut into 3-inch (7.5 cm) segments and bruised with the flat side of a knife

4 garlic cloves, minced

1-inch (2.5 cm) piece ginger, minced

½ serrano pepper, minced

½ pound (225 g) yardlong beans

1 pound (455 g) carrots, halved lengthwise and thinly sliced on the diagonal

1 cup (235 ml) coconut milk

1 tablespoon (15 ml) soy sauce

1 tablespoon (13 g) sugar

2 teaspoons ground turmeric

1 teaspoon ground cumin

½ teaspoon kosher salt

¼ teaspoon ground black pepper

Zest and juice of 1 lime

¼ pound (115 g) spinach, coarsely chopped

Handful of chopped cilantro for garnishing

Yardlong beans (also called Chinese long beans, asparagus beans, or snake beans) are thin, tight, and juicy pods that usually appear a little wrinkled, even when they're fresh. Their tender texture holds up well to a simmering curry dish in which every bite bursts with green bean flavor dressed in sweet coconut.

Throughout my childhood, yardlong beans used to grace the dinner table in spaghetti-like mounds. I'd always pick them up with my chopsticks and try to stuff an entire strand in my mouth, often with the ends dangling down my chin as I choked and giggled to my parents' dismay. This fondness for yardlong beans (which can grow up to 2 feet [60 cm] long) means I like to keep their length intact, but if you're more civilized than I am, you can cut them into more manageable pieces for this dish.

Heat a large sauté pan over medium-high heat. Swirl in the oil and add the onion, lemongrass, garlic, ginger, and serrano. Cook until the onion starts to turn translucent, 2 to 3 minutes. Add the beans and carrots, stir to coat with the aromatics, and cook for about 5 minutes. Add the coconut milk, soy sauce, sugar, turmeric, cumin, salt, pepper, lime zest, and lime juice, and bring to a boil.

Reduce the heat and simmer uncovered until the vegetables are tender, 8 to 10 minutes. Stir in the spinach and continue simmering until the leaves are wilted, 2 to 3 minutes. Discard the lemongrass before serving. Scatter a handful of chopped cilantro on top, if desired.

FAVA LEAF SALAD
WITH CITRUS, FETA, AND WALNUTS

MAKES 4 SERVINGS

for the dressing

1 tablespoon (10 g) minced shallot

½ teaspoon kosher salt

¼ teaspoon ground black pepper

2 teaspoons orange zest, plus more for garnishing

2 tablespoons (28 ml) olive oil

for the salad

4 cups (120 g) packed fava leaves

2 oranges, zested, segmented, and juices reserved (use zest for dressing)

2 grapefruits, segmented and juices reserved

¼ cup (38 g) crumbled feta cheese

½ cup (60 g) toasted and chopped walnuts

Fava flowers for garnishing

Weeks before the first fava beans appear in the garden or at the farmers' market, there's a surer sign of spring: fava greens. Many people overlook the silvery-green foliage in favor of the pods, but they're every bit as edible as the beans you're most familiar with. Fava greens are a great way to get that fresh fava flavor without all the work of shelling the beans. They're light, tender, succulent, and embody all the earthiness of spring.

If you're lucky enough to grow your own fava plants at home, or you come across the greens at your local farmstand, gather a few handfuls for this bright and zesty salad.

To make the dressing, combine the reserved orange juices and grapefruit juices with the shallot, salt, pepper, and orange zest in a small bowl. Slowly whisk in the oil until well blended.

Place the fava leaves in a large serving bowl and lightly toss with the dressing. (You might not use all of it.) Divide the fava leaves among four plates and arrange the orange and grapefruit segments around the leaves. Sprinkle the feta and walnuts on top, then garnish with a few fava flowers and a pinch of orange zest on each plate.

NO-STIR
FAVA BEAN RISOTTO

MAKES 4 TO 6 SERVINGS

5 cups (1.2 L) chicken broth

1 cup (100 g) chopped cauliflower

2 cups (300 g) shelled fava beans
(see Preparation Tip on page 97)

2 tablespoons (28 ml) olive oil

2 tablespoons (28 g) butter

1 yellow onion, finely chopped

2 garlic cloves, minced

1½ cups (270 g) uncooked arborio
rice

1 cup (235 ml) dry white wine

1 cup (100 g) grated Parmesan
cheese, plus more for serving

Lemon wedges for serving

COOK'S NOTE
A pound (455 g) of
fava bean pods yields
approximately ¾
cup (113 g) of twice-
shelled fava beans.

In all honesty, fava beans are a labor of love. Unless you have a houseful of kids you can put to work, or idle grandmothers rocking away on the porch, adding fava beans to the dinner menu always means an afternoon ritual at the kitchen table, freeing the beans from deep inside their bulky pods. But while favas really make you work for it, that first spoonful of rich, creamy risotto, paired with supple favas that taste of buttery spring, soon makes you forget why you were complaining.

Taste buds aside, what sets this risotto apart from many other risotto recipes is its (almost) no-stir nature. Instead of being tied to the stove for half an hour while you add cup after cup of broth, you can let it do its thing while it simmers away.

In a medium saucepan over medium-high heat, bring the broth to a boil. Add the cauliflower and cook until tender, 6 to 8 minutes. Using a slotted spoon, transfer the cauliflower to a blender along with 1 cup (235 ml) of broth and 1 cup (150 g) of fava beans and puree until smooth. Set the puree aside and reserve the remaining broth.

In a large, heavy sauté pan over medium-high heat, melt the butter in the oil. Add the onion and garlic and cook for 2 to 3 minutes until tender. Stir in the rice to coat. Pour in the wine and continue stirring until all of the liquid is absorbed, 2 to 3 minutes. Add the reserved broth and bring to a gentle boil.

Reduce the heat, cover, and simmer for 10 minutes. Give a quick stir and ensure the rice is evenly spread across the pan. Cover it again and continue simmering until most of the liquid is absorbed and the rice is pleasantly chewy, about 10 minutes.

Remove the lid and stir in the puree. Bring the risotto to a rapid boil and stir vigorously for 1 minute until thick and creamy. Be careful not to overcook it; when you tilt the pan, the rice should easily slide to the bottom without clumping together. Turn off the heat and stir in the remaining cup (150 g) of fava beans and Parmesan. Serve immediately with a squeeze of lemon and a dusting of more Parmesan on top.

1   2   3   4

PREPARATION TIP

1 Snap back the pointy end of the fava pod, toward the thicker seam, and pull down the string to "unzip" the pod.

2 Remove all the beans, which are covered in a waxy skin.

3 Blanch the beans until the skins dull in color, 2 to 3 minutes. Plunge them into an ice bath to stop the cooking.

4 When cool, use your fingers to pop the beans out of their skins.

CHARRED FAVA PODS
WITH PARMESAN

MAKES 4 SERVINGS

2 tablespoons (28 ml) olive oil

1 tablespoon (15 ml) balsamic vinegar

1 teaspoon kosher salt

¼ teaspoon ground black pepper

Zest and juice of ½ lemon

1 pound (455 g) fava beans

¼ cup (20 g) shaved Parmesan cheese

Here we were this whole time, made to think that in order to enjoy a fava bean, we had to shell it, blanch it, and shell it (again) before we can get to the good stuff. Not so! The secret to getting fava beans on the table quickly is to put them on the grill first. Over a hot grate, the pods turn tender and sweet, with a mildly smoky flavor, and even the waxy skins become edible.

The key is to look for medium-size beans that aren't overly bulging; once grilled, you can pop the whole pod in your mouth, skins and all. Even if your favas are more mature, grilling them softens the beans inside so you can open the pods, edamame style, and snack on the beans one by one.

Preheat the grill on medium-high heat.

In a large shallow dish, make a vinaigrette by combining the oil, vinegar, salt, pepper, and lemon juice. Add the fava beans and toss to coat. Arrange the fava beans in a single layer on the grill grate, cover, and grill until the pods are blistered on the bottom, about 5 minutes. Flip the fava beans and grill the other side for 2 to 3 minutes until the pods are softened and slightly charred and seem nearly ready to burst open.

Transfer the fava beans back to the dish, toss to coat with any remaining vinaigrette, then pile them onto a serving plate. Scatter the lemon zest and Parmesan on top before serving.

BULBS & STEMS

I have to say this is one of my favorite sections in the book, because it contains two members of the *Allium* genus, a group most widely known for its onions. And onions, if you haven't noticed by now, are one of my most-loved ingredients.

Leeks and scallions are nonbulbing onions, with the former resembling oversized green onions (though a leek is its own vegetable). And here's where it starts to get confusing: *green onion* is a general term for any young onion lacking a fully developed bulb, including spring onions and Mexican green onions, but it's more commonly known as a scallion in the United States. Sometimes *green onion* refers to the green leaves on an onion. And though you can pull a bunch of onions at the baby stage and call them scallions, they typically come from a crop called Welsh onions or Japanese bunching onions, which are bred to grow without a bulb.

I classify all of these alliums as stems, but botanically, the green tops of these plants are the leaves—so you can and should eat them, especially the leaves on leeks that are so often thrown away.

Fennel and kohlrabi are both considered bulb vegetables, but like the alliums above, their leaves are edible and delicious—if you can still find them attached to the bulbs, since they're never sold separately. They are a twofer of the plant world, with different parts that can be harvested at different times for a round-the-season crop. (And fennel is actually more than a twofer, since fennel pollen and fennel seeds are edible as well.)

FENNEL FROND AND
GINGER PESTO

MAKES 1 CUP (235 ML)

2 cups (18 g) packed fennel fronds

⅓ cup (33 g) toasted walnuts

1½ tablespoons (9 g) chopped ginger

1 garlic clove

¼ teaspoon kosher salt

Zest and juice of ½ lemon

¼ cup (60 ml) olive oil

Fennel fronds don't get as much respect as their bulbous counterparts do, but the feathery leaves share the same sweet flavor you'd expect from fennel. A typical fennel recipe will usually have you removing the fronds, and just a couple of bulbs will yield handfuls of leaves that are more than you need for garnishing. If you find yourself in that scenario, gather them up for this bright, gingery pesto that not only pairs well with a bowl of steaming pasta or a pan of roasted fish, it also adds zip to tropical fruit salads with mango or guava.

In a food processor, combine all the ingredients (except the oil) and pulse until crumbly, scraping down the sides of the bowl with a rubber spatula as needed. Continue pulsing and add the oil in a steady stream until smooth and well blended.

FENNEL AND SEAFOOD STEW

MAKES 6 SERVINGS

2 tablespoons (28 ml) olive oil

1 yellow onion, cut into small dice

1 fennel bulb, cut into small dice

4 garlic cloves, minced

2 tablespoons (32 g) tomato paste

¼ teaspoon ground black pepper

4 thyme sprigs

2 bay leaves

1 cup (235 ml) dry white wine

4 cups (946 ml) fish stock

1 pound (455 g) tomatoes, coarsely chopped

2 pounds (900 g) white fish fillets (such as cod, halibut, or tilapia), cut into 1-inch (2.5 cm) chunks

½ pound (225 g) uncooked large shrimp, peeled and deveined

¼ cup (15 g) chopped fresh parsley

Lemon wedges for serving

Chopped fennel fronds for garnishing

San Franciscans may find this stew somewhat reminiscent of their beloved cioppino, a seafood stew that originated in the "City by the Bay" in the late 1800s. Italian immigrants concocted this dish with leftovers from the day's catch, and over many generations, San Francisco chefs elevated this once-peasantly meal into a luxurious showcase of fish and shellfish.

I like to keep my own stew simple. The seafood tends to change, depending on what I find from the fishmonger (and oftentimes I'll mix the fish), though the base always remains the same: a savory tomato broth spiked with sweet fennel and heady herbs. Serve the stew with a crusty loaf of garlic bread and call it a (week)night.

Heat a large pot over medium-high heat. Add the oil, onion, fennel, and garlic, and cook until soft, 2 to 3 minutes. Add the tomato paste, pepper, thyme, and bay leaves and stir to coat thoroughly. Pour in the wine, bring the pot to a simmer, and cook until most of the liquid is evaporated, about 2 minutes. Add the stock and tomatoes and bring to a boil. Once the stock reaches a consistent boil, reduce the heat and simmer uncovered for 20 minutes to let the flavors meld together.

Gently stir in the fish and simmer for 3 minutes. Stir in the shrimp and cook until the shrimp is opaque and the fish is flaky, 3 to 5 minutes. Remove and discard the bay leaves. Stir in the parsley. Serve immediately with a squeeze of lemon over each bowl and a sprinkle of fennel fronds, if desired.

COOK'S NOTE
If you can't find good-quality fish or seafood stock for the stew, you can cheat by using vegetable broth (or half-vegetable and half-chicken broth for a stronger flavor) and adding a few spoonfuls of bottled clam juice or fish sauce to taste.

CARAMELIZED
FENNEL AND ONION

MAKES 4 SERVINGS

1 fennel bulb, sliced lengthwise into 1-inch (2.5 cm) wedges

1 yellow onion, sliced lengthwise into 1-inch (2.5 cm) wedges

1 tablespoon (15 ml) olive oil

½ teaspoon kosher salt

¼ teaspoon ground black pepper

Many people shy away from fennel, which they describe as having too strong of a licorice-like flavor. When you put that same fennel in the oven under high heat, however, its love-it-or-hate-it aroma mellows out into a warm slice of sweetness. Fennel bulb caramelizes beautifully the way onion does, turning soft and fragrant with only the slightest hint of anise. After a long roast, the sumptuous flavors of fennel and onion marry to make a deep, rich, and smoky sweet side to a savory steak.

Preheat the oven to 425°F (220°C, or gas mark 7).

On a large rimmed baking sheet, toss the fennel and onion with the oil, salt, and pepper until thoroughly coated. Scatter the vegetables across the baking sheet in a single layer and roast until golden brown and slightly charred on the edges, 35 to 40 minutes. Halfway through the roast, give the fennel and onion a quick stir for even caramelization on all sides.

KOHLRABI AND CARROT SLAW

MAKES 6 SERVINGS

for the slaw

1 pound (455 g) kohlrabi, peeled and julienned or cut into 4-inch (10 cm) matchsticks

½ pound (225 g) carrots, peeled and julienned or cut into 4-inch (10 cm) matchsticks

1 teaspoon kosher salt

¼ red onion, thinly sliced

¼ cup (60 ml) chopped cilantro

½ jalapeño pepper, minced

for the dressing

⅓ cup (75 g) mayonnaise

2 tablespoons (30 g) Dijon mustard

⅛ teaspoon ground black pepper

Zest and juice of ½ lime

I take a savory direction with this slaw, making it bold and creamy rather than sweet and tangy. It holds its own as a standalone salad, but I especially like it over tacos or heaped on a bowl of carnitas. As you might imagine, this spicy slaw—tempered with a touch of acidity from the lime—pairs well with rich, fatty meats, especially if they've been singed on the grill.

In a large colander, toss the kohlrabi and carrots with the salt and let drain in the sink for 1 hour. Periodically shake up the vegetables to expel as much water as possible. The strips should be tender and pliable but still retain their crispness.

Meanwhile, make the dressing by combining all of its ingredients in a small bowl. Set aside until ready to use.

In a large serving bowl, combine the kohlrabi, carrots, onion, cilantro, and jalapeño. Toss with the dressing to coat and refrigerate for at least 1 hour before serving. Serve chilled.

KOHLRABI HOME FRIES
WITH THYME AIOLI

MAKES 4 SERVINGS

for the fries

2 pounds (900 g) kohlrabi, peeled and cut into 3-inch (7.5 cm) spears

2 tablespoons (28 ml) olive oil

½ teaspoon kosher salt

¼ teaspoon garlic powder

⅛ teaspoon ground black pepper

for the aioli

1 egg

2 garlic cloves, crushed

1 tablespoon (2 g) chopped fresh thyme

1 teaspoon lemon juice

¼ teaspoon kosher salt

¾ cup (175 ml) sunflower oil

At first sight, kohlrabi can seem like an intimidating thing. The orb is often green and pale but also comes in a shocking purple, and its tentacle-like stems are reminiscent of Sputnik 1. But underneath that slightly rubbery and sometimes knobby exterior is a smooth, white, and firm flesh that turns tender-crisp in a hot oven, cooking up like a lighter version of potato fries. They have every bit of the satisfying munch of a spud, but leave you with none of the heaviness in your belly.

Preheat the oven to 425°F (220°C, or gas mark 7).

On a large rimmed baking sheet, toss the kohlrabi with the oil, salt, garlic powder, and pepper. Scatter the kohlrabi across the baking sheet in a single layer and bake for 35 to 40 minutes until lightly browned, shaking them up halfway through to evenly brown all the sides.

Meanwhile, make the aioli by adding the egg, garlic, thyme, lemon juice, and salt to a blender. Blend on medium speed for a few seconds until well combined. While the blender is running, add the oil in a very slow, steady, and thin (think needle-size) stream until the mixture emulsifies. Don't try to rush the stream of oil; the emulsification starts slowly, but you'll hear the sound of the motor change as the aioli thickens and starts slapping the sides of the blender. When the aioli turns opaque and smooth, transfer to a small bowl and serve with the kohlrabi fries.

COOK'S NOTE
I use sunflower oil to emulsify the aioli, but you can substitute another neutral oil, such as safflower or grapeseed. I don't recommend using extra-virgin olive oil with any aioli made in a blender or food processor, as the high-speed action of the whirring blades can bring out its bitterness. Extra-virgin olive oil can also cause an emulsification to break more easily.

SHORTCUT TO PERFECT AIOLI

If you have an immersion blender, there's an easier way to perfect aioli. Simply add all of the aioli ingredients to a blending cup (preferably the one that came with your immersion blender) or a mason jar. The oil will naturally separate and float to the top. Press the immersion blender down into the cup and blend on high speed for about 1 minute, gradually moving the blender up the cup as the mixture starts to emulsify. Voilà, effortless aioli!

KOHLRABI GREEN AND WILD MUSHROOM RAGOÛT
WITH POLENTA

MAKES 4 SERVINGS

for the polenta

Butter or oil for greasing

3¼ cups (760 ml) lukewarm water

1 cup (140 g) coarse yellow cornmeal (polenta)

1 tablespoon (14 g) butter, softened

1 teaspoon salt

for the ragoût

½ ounce (15 g) dried porcini mushrooms

1 cup (235 ml) hot water

2 tablespoons (28 g) butter

2 tablespoons (28 ml) olive oil

8 ounces (225 g) mixed fresh wild mushrooms, coarsely chopped

½ teaspoon kosher salt

⅛ teaspoon ground black pepper

2 shallots, chopped

2 garlic cloves, chopped

1 teaspoon all-purpose flour

2 cups (134 g) chopped kohlrabi greens

½ cup (120 ml) dry red wine

1 teaspoon chopped fresh thyme

When it comes to root vegetables, kohlrabi bulbs are as esoteric as they come. But kohlrabi greens are nearly mythical; unless you grow them in your garden, they're seen only when the farmer or the grocer hasn't stripped the bulbs of their cabbagelike foliage. The leaves sprout from tentacle-like stems on the bulbs and true to their lineage, cook up just like cabbage, kale, collard greens, and broccoli greens. (In fact, you can substitute any of these cruciferous vegetables if you don't have kohlrabi greens for this recipe.) The leaves turn silky and savory in a sauté with wild mushrooms, which are then piled on a bed of buttery polenta for this vegetarian stew.

Preheat the oven to 350°F (180°C, or gas mark 4).

Combine all of the polenta ingredients in a greased 8-inch (20 cm) square baking dish (don't worry about getting the butter to melt right away) and bake for 50 minutes. Stir up the polenta a bit with a fork and continue baking for 10 minutes.

Meanwhile, rehydrate the porcini mushrooms by soaking them in the hot water for 20 minutes. Drain the mushrooms, reserving all of the mushroom stock, and squeeze out any excess water.

In a large skillet over medium-high heat, melt the butter in the oil. Add the fresh wild mushrooms, salt, and pepper, stir to coat with oil, and cook until they release their liquid and become tender, 6 to 8 minutes. Stir in the shallots and garlic and cook until the shallots turn translucent, about 5 minutes. Sprinkle the flour on top and stir to combine.

Increase the heat to high. Stir in the porcini mushrooms, kohlrabi greens, wine, and thyme, and bring to a boil. Cook until the liquid reduces slightly and glazes the vegetables, 1 to 2 minutes. Pour in the reserved mushroom stock and bring to a rapid boil for 1 minute. Reduce the heat and simmer uncovered for about 5 minutes, stirring occasionally until the liquid thickens and the vegetables are tender. Serve the ragoût over a bowl of warm polenta.

LEEK GREEN, WILD MUSHROOM, AND
GOAT CHEESE CROSTINI

**MAKES 8 TO 10
SERVINGS**

1 baguette (12 inches, or 30 cm), sliced ¼ inch (6 mm) thick on the diagonal

3 tablespoons (45 ml) olive oil, divided

2 garlic cloves, minced

2 cups (178 g) finely chopped leeks (dark green leaves only, see Preparation Tip below)

½ pound (225 g) mixed wild mushrooms, finely chopped

½ teaspoon salt

⅛ teaspoon ground black pepper

⅓ cup (80 ml) dry white wine

¼ cup (15 g) chopped fresh parsley

4 ounces (115 g) creamy goat cheese (chèvre)

Most recipes tell you to use the white parts of the leeks and to discard the green parts, and over time, this has led people to believe the green parts are inedible. Their leathery appearance suggests they might be tough and fibrous, but underneath the sometimes dirty leaves and wilted ends are the tender tops I consider a high point of leeks, literally and figuratively.

A simple sauté with wild mushrooms and a simmer in white wine brings out the characteristic sweetness of leeks, turning them soft and fragrant. Pair them with pungent goat cheese on a toasted baguette, and you've got yourself a party-worthy crowd-pleasing appetizer.

Preheat the oven to 375°F (190°C, or gas mark 5). Spread the baguette slices across a large, rimmed baking sheet in a single layer and drizzle 2 tablespoons (28 ml) of oil over them. Bake until crispy and golden brown, 8 to 10 minutes.

Meanwhile, heat a large skillet over medium-high heat. Swirl in the remaining 1 tablespoon (15 ml) of oil and add the garlic and leeks. Stir and cook until the leeks are tender, about 5 minutes. Add the mushrooms, salt, and pepper, and cook until the mushrooms are soft and any juices have evaporated, 6 to 8 minutes. Pour in the wine and bring the mixture to a rapid boil for 1 minute, then reduce the heat and simmer until all of the liquid is absorbed. Turn off the heat and scatter parsley across the top.

To assemble the crostini, smear some goat cheese on a slice of toasted baguette and top with the leeks and mushrooms. Serve warm or at room temperature.

PREPARATION TIP

The key to using leek tops is to cut the green leaves off the white stem and wash them separately. As the part of the plant that grows above ground, leek tops gather even more grit than the bottoms and it takes a good scrubbing to make sure you get all the dirt between the layers and inside the creases. If the leaves look especially beaten up, remove and discard the outermost layer; often you'll find fresher and more tender leaves tucked inside.

LEEK AND BACON
BREAKFAST PIZZA

MAKES 4 SERVINGS

4 leeks (white stems and dark green leaves)

4 bacon strips

4 whole-grain pitas

Olive oil for brushing

1½ cups (173 g) shredded Monterey Jack cheese

4 eggs

Freshly cracked black pepper to taste

I could eat breakfast pizza for breakfast, lunch, and dinner. And while you might be thinking that pizza in the morning sounds like a labored undertaking, hear me out: This kind of pizza comes together in less than half an hour, and there's no kneading or rolling in sight.

I use whole-grain pitas as personal-size crusts, and they bake up so thin and crispy, you could be fooled into thinking they were made for pizza. Sweet and silky leeks don't make you miss the sauce at all, and melted Monterey Jack makes a creamy pillow for crunchy bits of bacon. But, perhaps my favorite part of all is slicing into a piping-hot pizza and letting the luscious yolk ooze over the toppings. If you're not a fan of runny yolks, simply bake for an extra minute or two.

Preheat the oven to 425°F (220°C, or gas mark 7).

To prepare the leeks, trim the root ends and cut the leeks in half where the white stems and green leaves meet. Slice the white stems in half lengthwise and wash both sides thoroughly to remove any trapped dirt. Then, slice crosswise into thin half moons and set aside. Rinse the green leaves, making sure to wash all the grit between the layers, and thinly slice them crosswise. Set aside and keep the greens separate from the whites.

Heat a large skillet over high heat. Fry the bacon for about 3 minutes, then flip and fry for about 2 minutes until crisp to your liking. Drain the bacon on paper towels, reserving all the bacon grease in the skillet.

Reduce the heat to medium. Add the green leeks to the skillet and cook until the greens start to wilt, about 5 minutes. Add the white leeks and cook for 5 minutes, stirring occasionally, until the whites are translucent and all the leeks are soft.

Place the pitas on two large rimmed baking sheets. (If your oven cannot fit two baking sheets side by side, bake the pitas in two batches.) Brush the pitas generously with the oil, making sure to coat the edges. Chop the bacon, then top each pita with an equal heaping of leeks, cheese, and bacon.

In the center of each pita, make a small well with a spoon and crack an egg into it. Bake until the egg whites are set but the yolks are still slightly runny, 10 to 12 minutes. Be sure to keep an eye on your oven so the eggs don't overcook. Serve with a few turns of freshly cracked black pepper on top.

LEEK GREEN AND SAUCY
SHRIMP STIR-FRY

MAKES 4 SERVINGS

for the sauce

¼ cup (60 ml) soy sauce

1 tablespoon (15 ml) sriracha

1 tablespoon (15 ml) rice vinegar

1 tablespoon (15 ml) sesame oil

1 teaspoon sugar

for the stir-fry

2 tablespoons (28 ml) sunflower oil

4 cups (356 g) chopped leeks (dark green leaves only, see Preparation Tip on page 112)

4 garlic cloves, minced

1-inch (2.5 cm) piece ginger, minced

1¼ pounds (570 g) uncooked large shrimp, peeled and deveined

I always cringe when I hear people say they cut all the leaves off their leeks before starting a recipe. They are throwing away perfectly good food and don't even know it. This recipe *starts* with the leaves, the fan-shaped, dark green tops of the white stems most people associate with leeks.

Though they look a little tougher than their white counterparts, the leaves cook down just as tenderly as the stems and share the same sweet, buttery flavor. Sometimes I actually prefer the leaves to the stems, as their texture won't break down under high heat. They end up caramelizing much like onions do, turning soft and velvety but retaining a bit of bite. Their mildness makes them a clean complement to the bold, spicy sauce in this stir-fry.

In a small bowl, combine all of the sauce ingredients and set aside.

Heat a wok or large skillet over medium-high heat. Swirl the sunflower oil around the wok and add the leeks, spreading them across the surface in a single layer. Cook undisturbed until they begin to wilt and brown on the edges, about 2 minutes. Give a quick stir and continue cooking for 2 to 3 minutes until the leeks are tender. Add the garlic and ginger, tossing them around frequently so they don't burn.

Add the shrimp and stir-fry until the flesh begins to turn pink on both sides, 3 to 4 minutes. Pour in the sauce and toss to coat. Remove from the heat and serve immediately.

SEARED SCALLOPS OVER MELTED LEEKS
WITH TARRAGON BUTTER SAUCE

MAKES 4 SERVINGS

for the leeks

6 leeks (white and light green stems only)

2 tablespoons (28 g) butter

¾ cup (175 ml) chicken broth

½ teaspoon kosher salt

¼ teaspoon ground black pepper

for the scallops

1¼ pounds (570 g) sea scallops, rinsed

¼ teaspoon kosher salt

⅛ teaspoon ground black pepper

1 tablespoon (14 g) butter

for the sauce

½ cup (120 ml) dry white wine

Juice of ½ lemon

4 tablespoons (55 g) butter

2 tablespoons (8 g) minced tarragon

Chopped fresh parsley for garnishing

Lemon zest for garnishing

What exactly are melted leeks? They're sweet, fragrant leeks, braised on low heat over a long period to the point where they're so tender, they nearly fall apart when you pick them up with a fork. At that first bite, the luxurious leeks seem to melt on your tongue.

When it comes to perfect pairings, leeks and scallops go hand in hand. Leeks don't overpower the mildness of butter-seared scallops, and they soak up a rich tarragon butter sauce that adds a lovely aroma to the dish. Don't be afraid of all that butter. There's a lot of butter in this recipe once you add it all up, but it's a nice treat.

To prepare the leeks, slice the stems in half lengthwise and rinse under running water, fanning out the layers to wash away the grit. Shake dry and slice the leeks crosswise into thin half moons.

Melt the butter in a large skillet over medium-low heat and stir in the leeks to coat. Add the broth, salt, and pepper, cover, and cook until the leeks are soft, 20 to 25 minutes. Uncover and let any remaining broth cook off. Give a stir, then reduce the heat to low and keep them on the stove while you make the scallops and tarragon butter sauce.

To prepare the scallops, thoroughly pat them dry with paper towels and sprinkle with the salt and pepper. Melt the butter in a large sauté pan over medium-high heat and swirl it around to coat the bottom of the pan. Place the scallops in a single layer in the hot pan and sear on one side, without moving them, for 3 minutes. Flip the scallops and sear them on the other side until golden brown, 2 minutes. With scallops, it's best to err on the side of undercooking them a bit as they'll continue to cook once they're off the heat. Slightly undercooked scallops are still delicious, while slightly overcooked scallops become rubbery. As soon as the scallops start to turn opaque throughout, transfer them to a platter and keep warm.

To make the tarragon butter sauce, pour the wine and lemon juice into a small saucepan over medium-high heat and bring to a boil. Boil until the liquid is reduced in half, 3 to 5 minutes. Turn off the heat and whisk in 2 tablespoons (28 g) of the butter until softened but not completely melted.

Whisk in the remaining 2 tablespoons (28 g) of butter and the tarragon until the sauce is smooth and thick.

To serve, divide the leeks among four plates and top with the scallops. Drizzle the tarragon butter sauce over the scallops and scatter the chopped parsley and lemon zest on top, if desired.

THE SECRET TO GETTING A GREAT SEAR

To plump up, preserve, and whiten them, many scallops are soaked in a phosphate solution for selling. They're labeled as "wet" scallops and the phosphates not only prevent a crispy brown crust, they also give a slightly soapy taste. Wet scallops (which tend to be snow white in color) absorb much of this liquid, so often you're paying for all the water that's been added to their weight. Whenever possible, buy scallops labeled "dry," "dry packed," or "chemical free," which are higher quality. Dry scallops tend to be more ivory in color, and since they're not waterlogged, they get a great sear in the pan.

GRILLED GREEN ONIONS
WITH CHILE LIME MARINADE

MAKES 4 SERVINGS

¼ cup (60 ml) olive oil

⅓ cup (5 g) chopped cilantro, plus more for garnishing

1 jalapeño pepper, minced

2 garlic cloves, minced

Zest and juice of 2 limes

½ teaspoon ground cumin

½ teaspoon kosher salt

1 pound (455 g) green onions (scallions), ends trimmed

For a lazy weeknight dinner, the grill is where it's at. Less time is required, less cleanup is needed, and fewer dishes are involved, all of which are a win-win-win in my book. It usually starts with a steak, and some of my favorite accompaniments for a juicy slab are these lightly charred and caramelized green onions. The stems turn silky and smoky over an open flame, and they go from grill to table in the time it takes for your steaks to rest.

In a small bowl, combine all of the ingredients (except the green onions) to make a marinade. Place the green onions in a large shallow dish and toss with the marinade to coat. Let stand at room temperature for at least 15 minutes for the flavors to meld together.

Meanwhile, preheat the grill on medium-high heat. Arrange the green onions across the grill grate in a single layer and grill for 5 to 6 minutes, turning once or twice until the stems are softened and slightly charred. Transfer to a serving dish and garnish with chopped cilantro, if desired.

COOK'S NOTE
Try to find thicker-stemmed scallions so they retain a bit of bite after cooking.

GREEN ONION PANCAKE
WITH SPICY SOY DIPPING SAUCE

MAKES 4 SERVINGS

for the dipping sauce

¼ cup (60 ml) soy sauce

2 tablespoons (28 ml) rice vinegar

1 teaspoon sugar

1 teaspoon toasted sesame seeds

½ serrano pepper, thinly sliced

for the pancakes

2 cups (250 g) all-purpose flour

1½ teaspoons kosher salt

2 cups (475 ml) cold water

¼ cup (60 ml) sunflower oil, divided

½ pound (225 g) garlic chives, ends trimmed and stems cut into 5- to 6-inch (13 to 15 cm) lengths (reserve 2 full stems for dipping sauce)

Vietnamese, Chinese, Korean, and Japanese cultures all have their own versions of this savory pancake, often made with whole garlic chives. These are not the onion chives typically used as garnish. Garlic chives are used as a vegetable in Asian cooking, and true to their name, have a subtle garlic flavor. If you're one to love the smell of onion and garlic wafting through the kitchen as you cook (is there anything more inviting?), these rustic pancakes deserve a place on your dinner menu.

To make the dipping sauce, combine all of its ingredients in a small bowl. Thinly slice the reserved chives and stir them into the sauce. Set aside.

In a medium bowl, combine the flour and salt. Whisk in the water until the batter is well blended and free of lumps.

Heat a medium skillet over medium heat and coat the surface evenly with 1 tablespoon (15 ml) of the oil. Spread one-quarter of the chives across the skillet in a single layer, then pour 1 cup (235 ml) of the batter evenly over the chives. (Don't worry about making a perfect circle of batter like a traditional pancake; this savory version is meant to have a rustic and free-form shape. The batter is more about holding the chives together and less about forming a pancake.) Fry until the edges of the pancake start to crisp and the bottom is golden brown, about 4 minutes. Gently flip the pancake and fry the other side until cooked through, about 2 minutes. The pancake should feel springy and have a slightly chewy texture. Transfer to a serving dish and cover, then repeat the process with the remaining three pancakes. Serve warm with the dipping sauce.

THE GREEN ONION GROUP

Garlic chives are nonbulbing green onions that go by a smattering of other
names, such as Asian chives, Chinese chives, Chinese leeks, or Oriental garlic
(and though they're all part of the same family of green onions, they're not the
kind of leeks or garlic most of us are familiar with). Garlic chives are long and
slender like scallions, about 12 inches (30 mm) in length, but have thin and
flattened leaves. Sometimes they'll come with flower buds on the ends, which
are also edible and delicious. For this recipe, you can substitute thin stems of
scallions, ramps, or green garlic, but their texture is best if thinly sliced.

OYAKODON-STYLE OMELETTE
WITH SPRING ONIONS

MAKES 4 TO 6 SERVINGS

1 tablespoon (15 ml) olive oil

1 pound (455 g) spring onions, quartered lengthwise

1 pound (455 g) boneless skinless chicken thighs, cut into ½-inch (1.3 cm) pieces

4 cremini mushrooms, quartered lengthwise

⅓ cup (80 ml) chicken broth

⅓ cup (80 ml) soy sauce

⅓ cup (80 ml) dry white wine

1 teaspoon sugar

8 eggs

¼ cup (25 g) thinly sliced scallions

1 cup (55 g) packed baby greens

COOK'S NOTE
Look for spring onions no larger than 2 inches (5 cm) in diameter. If you can find a bunch that comes with fresh, healthy green stems, slice up those stems in place of the scallions.

This is one of those breakfast-for-dinner kinds of meals—quick, comforting, and nourishing. When I used to live in downtown Los Angeles, I'd order *donburi* (rice bowls—sometimes shortened to just *don*, like in *oyakodon*) at my favorite late-night Japanese joint. It would be 2 a.m., tables packed with club kids and even more waiting by the door, and I'd eagerly sit at the grimy counters coated in wok grease just to get my fill of those steaming bowls of eggs and rice. After moving many miles away and searching for some midnight grub to satisfy my umami craving, I learned how to make my own version of oyakodon at home.

Oyakodon is a Japanese home-style meal that means, quite literally, "parents and kids bowl," and it's a playful pun on the fact that both chicken and egg are used in the dish. (In this case, the chicken comes first.) Traditionally it's heaped atop a bowl of steamed rice to soak up all the sauce, but it's certainly hearty without it, too.

Heat a large skillet over medium-high heat. Add the oil and onions and cook for about 5 minutes until tender. Add the chicken, mushrooms, broth, soy sauce, wine, and sugar and stir to combine. Bring the liquid to a boil, then reduce the heat and simmer uncovered, stirring occasionally until the chicken is cooked through and no longer pink, about 5 minutes.

In a medium bowl, very lightly beat the eggs for just a few seconds, leaving some of the yolks slightly intact. Pour half the eggs evenly into the skillet and cook undisturbed until they barely begin to set, about 3 minutes. Pour the remaining eggs into the skillet, cover, and cook until the omelette is thickened, about 2 minutes. Turn off the heat and allow the bubbling to subside a bit. Scatter the scallions and baby greens on top and serve warm in the skillet.

6

ROOTS & TUBERS

Roots and tubers sometimes seem like the underdogs of the vegetable world. No matter how hard-working they are, how many roasts they perfume or last-minute potlucks they save, these dirty, hairy, gnarled, stout pieces of flesh are never appreciated in their raw, real form. We often vigorously scrub them, wash them, and peel them before we present them. Yet aside from beets—which I always peel, as the skin tends to be too gritty—I love the rustic quality of unpeeled vegetables, from carrots and radishes to potatoes and sweet potatoes.

What's a roasted carrot without that perfectly caramelized wrinkled carrot skin? Or a French fry without the satisfying crispness of potato skin?

Besides, science says eating dirt is good for you. This is not to say we should start making mud pies; but the specks of dirt on the skins of vegetables are generally harmless, especially if they're grown organically. In fact, the flora found in soil can actually help build beneficial bacterial colonies in our guts, protecting us from toxins, parasites, and pathogens. So ease up on the scrubbing and embrace the skins on these vegetables!

With the exception of potatoes (no relation to sweet potatoes), the tops of these roots and tubers are edible as well. Beet greens and sweet potato vines turn soft and sweet in a sauté, while carrot tops and radish greens are good eatin' in salads. This is a food group where waste is not— *should* not—be an issue, ever.

SKILLET GREENS AND BACON BITS
WITH POMEGRANATE GASTRIQUE

MAKES 4 TO 6 SERVINGS

for the gastrique

¾ cup (175 ml) pomegranate juice

¼ cup (60 ml) red wine vinegar

¼ cup (50 g) sugar

for the skillet greens

3 bacon strips

2 shallots, thinly sliced

1 teaspoon red pepper flakes

1½ pounds (680 g) beet greens, leaves sliced into 1-inch (2.5 cm) strips crosswise and stems cut into 1-inch (2.5 cm) pieces, set aside separately

1 cup (174 g) pomegranate seeds

As you might have guessed from my Down-Home Collard Greens (page 72), I always love me a big heaping pile of greens. These ones are a bit more gourmet (or as I like to call it, gour-mette) though, as the greens are splashed with a pomegranate gastrique before serving.

A gastrique is simply a sauce made from a reduction of vinegar and sugar—and in this case, pomegranate juice is added for a bold, forward flavor. It takes only minutes to make, but gives fancy flair to a down-home skillet. The balance of sweet, salty, and spicy in this dish will have all your taste buds high-fiving each other.

To turn this dish into a full meal, grill up your favorite steak and drizzle some of the pomegranate gastrique over it. This dish also pairs well with pork, chicken, duck, rice, and potatoes.

In a small saucepan over medium-high heat, combine all the ingredients for the gastrique and bring to a boil. Maintain a gentle boil, stirring occasionally until the liquid has reduced by nearly half, 7 to 8 minutes. Keep warm and set aside.

Heat a large skillet over high heat. Fry the bacon for about 3 minutes, then flip and fry for about 2 minutes until crisp to your liking. Drain the bacon on paper towels and reserve all the bacon grease in the skillet.

Reduce the heat to medium-high and add the shallots and red pepper flakes. Stir and cook until fragrant, about 1 minute. Add the beet stems and cook until they start to become tender, about 5 minutes. Toss in a couple of handfuls of beet leaves, turning frequently to coat all the leaves in bacon grease. As soon as those leaves start to wilt, push them aside with a spatula and add another couple of handfuls of leaves; repeat until they're all in the skillet. Cook until the stems and leaves are tender, 15 to 20 minutes. Pour in the gastrique and toss to coat the leaves evenly. (You might not need to use all of it.) Gently fold in the pomegranate seeds, then remove the skillet from the heat.

Crumble the bacon on top. If you have any gastrique left over, serve it at the table for dipping or pouring over your other dishes.

BEETZA BEETZA

MAKES 4 SERVINGS

for the crust

2¼ teaspoons (1 packet) active dry yeast

1 teaspoon sugar

1 cup (235 ml) warm water (about 100°F to 110°F [38 to 43°C])

2½ cups (313 g) all-purpose flour

1 teaspoon kosher salt

2 tablespoons (28 ml) olive oil, plus more for greasing and brushing

for the toppings

3 beets, peeled and sliced into ¼-inch (6 mm) rounds

1 tablespoon (15 ml) olive oil

¼ teaspoon kosher salt

A few grinds of black pepper

10 beet leaves, finely chopped

1 shallot, thinly sliced

⅓ cup (40 g) crumbled blue cheese

⅓ cup (38 g) shredded mozzarella cheese

COOK'S NOTE
If you prefer a thin-crust pizza, you can split the dough and freeze the other half. Roll out the remaining dough per the instructions.

Homemade pizza on a weeknight? Without all the kneading, the resting, the rising, and the resting again? Yes, it's possible. This same-day pizza dough nearly changed my life! Now I don't even need to *think* about making pizza dough a day ahead of time. I can just mix it up when I get that craving, and it's ready by the time the toppings are ready. It yields a no-fuss crust that's chewy and soft and lends itself to lots of variations if you want more flavor (add Italian seasonings to the dough), more crispness (use a pizza stone to bake), or more meal options (try it for breadsticks and calzones, too).

This sauceless pizza came together one night when I didn't have much in the fridge, but I did have some beautiful beets. The combination of sweet juicy beets, earthy tender beet greens, and salty creamy blue cheese soon earned its place on my list of easy (and impromptu) weeknight dinners.

Preheat the oven to 450°F (230°C, or gas mark 8).

In a medium bowl, dissolve the yeast and sugar in the warm water and let stand until the surface forms a creamy layer, about 10 minutes. Stir in the flour, salt, and oil and mix with a wide spoon until the flour is incorporated and no dry pockets remain. Loosely form the dough into a ball with the spoon and let it rest for 20 minutes while you prepare and roast the beets.

In a small baking dish, toss the beets with the oil, salt, and pepper. Cover the dish with foil and roast until the beets are fork-tender, 15 to 20 minutes. Remove the beets from the oven, but leave the heat on.

Turn the dough onto a floured surface and roll into a 12-inch (30 cm) round. Transfer to a preheated pizza stone, greased pizza pan, or a greased unrimmed baking sheet and par-bake for about 5 minutes until puffed and crispy. Remove the crust from the oven and layer the beet leaves, shallot, blue cheese, and beets on top, leaving a 1-inch (2.5 cm) border all around. Sprinkle mozzarella over the toppings and brush a thin layer of oil over the crust. Finish baking the pizza in the oven until the cheese is melted and the crust is golden brown, about 15 minutes.

SHAVED RAW BEET SALAD
WITH WARM PECAN DRESSING

MAKES 4 SERVINGS

for the dressing

¼ cup (60 ml) olive oil

¼ shallot, minced

2 garlic cloves, minced

¼ cup (60 ml) cider vinegar

2 teaspoons honey

¼ cup (28 g) toasted and chopped pecans

Pinch of kosher salt

A few grinds of black pepper

for the salad

2 Chioggia beets, peeled and thinly sliced

2 cups (76 g) thinly sliced beet greens, stems removed

Crumbled feta cheese for serving

When it comes to root vegetables, I love finding ways to put the tops and bottoms back together in my recipes (like my Butter-Braised Radishes and Radish Greens with Farro, page 145). This salad is another one of those concoctions, a meal that makes use of the entire vegetable you bring home from the market or pull from your garden.

Beet recipes usually call for roasting the roots to bring out their sweetness, but raw beets are actually quite sweet on their own, in an earthy kind of way. Here, they're paired with a warm honey-infused dressing to heighten their natural flavor. Use a mandoline to shave off paper-thin slices of candy-striped Chioggia beets and toss them with tender beet greens for a conversation-starting salad. It's wonderful with baby beets, too. (Which way you slice them is up to you: crosswise for concentric circles or lengthwise for zebralike stripes.)

To make the dressing, combine the oil, shallot, and garlic in a small saucepan over medium heat. Cook until fragrant, about 5 minutes. Whisk in the vinegar and honey until well blended, then add the pecans, salt, and pepper. Stir to combine and keep warm.

In a large serving bowl, toss the beets and beet greens with the warm dressing. Serve with a sprinkle of feta on top.

ROSEMARY-ROASTED
CARROTS ON A PILLOW OF PESTO

MAKES 4 SERVINGS

1½ pounds (680 g) carrots

2 tablespoons (28 ml) olive oil

1 tablespoon (2 g) chopped fresh rosemary

½ teaspoon kosher salt

1 cup (260 g) pesto (homemade or store-bought)

This is a simple side of roasted carrots. What's not to like? Especially if you turn it into an elegant serving with a generous swipe of pesto on the plate—*then* it looks a little fancier for company. This is why I like to keep a jar of pesto around for impromptu meals; it can elevate even the most basic preparation into something presentation-worthy. Sweet caramelized carrots also happen to be the perfect vehicle for a fresh herbal pesto.

Try this recipe with Fennel Frond and Ginger Pesto (page 103) for a warm, fragrant flavor.

Preheat the oven to 425°F (220°C, or gas mark 7).

On a large rimmed baking sheet, toss the carrots with the oil, rosemary, and salt until evenly coated. Spread the carrots across the baking sheet in a single layer, then roast until the flesh is fork-tender and the tops are slightly browned, 15 to 20 minutes.

Spread the pesto across a serving platter with a large spoon, forming a thick pillow for the carrots. Pile the carrots on top and serve. You can also divide this dish among four smaller plates for individual servings.

CARROT TOP SALSA

**MAKES 2 CUPS
(475 ML)**

2 cups (120 g) minced carrot greens (leaves and tender stems only)

3 tablespoons (30 g) minced garlic

3 tablespoons (12 g) minced fresh oregano

2 tablespoons (11 g) minced jalapeño pepper

1 to 1¼ cups (235 to 285 ml) olive oil

¼ cup (60 ml) red wine vinegar

Zest and juice of 1 lemon

Carrot tops are those feathery fronds that most people—if they get them at all—pull off as soon as they bring their carrots home. Sometimes they go straight in the trash. Or the compost. Or the rabbit cage. But did you know that carrot tops make a fine substitute for parsley and even qualify as a leaf vegetable on their own? Their slightly chewy texture makes them best suited for finely chopping up into recipes, but they add a unique, earthy, carrot-like flavor to salads, soups, and even this salsa.

If I have a bunch of carrot tops left over from a carrot harvest, I'll make this tangy condiment. Use it as a dip for chips and bread, a marinade for meats and vegetables, or a drizzle for omelettes and roasts. (I'm sure you'll find plenty of other uses for it, too!)

Add all of the ingredients to a medium bowl and stir to combine. (Use more or less oil as desired, to make a chunkier or a thinner sauce.) Cover and let stand at room temperature overnight while the flavors intermingle. Carrot top salsa only gets better with age, so you'll know it's good when the carrot tops have turned a deep, muted shade of army green.

Decant the salsa into a jar and refrigerate. The oil may congeal in the cold temperature, but this will not affect the flavor. Bring the salsa to room temperature before serving.

VIETNAMESE CARROT AND DAIKON PICKLES

MAKES 3 CUPS (700 ML)

½ pound (225 g) carrots, cut into 2-inch (5 cm) matchsticks

½ pound (225 g) daikon, cut into 2-inch (5 cm) matchsticks

2 teaspoons kosher salt

1 cup (235 ml) rice vinegar

1 cup (235 ml) water

¼ cup (50 g) sugar

When I was growing up, a hefty jar of Vietnamese carrot and daikon pickles (*đồ chua* to us folk) were as ubiquitous in our kitchen as kosher dill pickles were in my friends' kitchens. We heaped them onto noodle bowls (*bún thịt nướng*), rice dishes (*cơm tấm*), and baguettes (*bánh mì*)—any dish that could benefit from a little sweet and sour crunch. I'd even sneak fingerfuls out of the jar to eat them as a snack!

As I grew older, I started using *đồ chua* in non-Vietnamese dishes as well, piling the pickles on hamburgers and hot dogs, grilled chicken wraps, and mixed green salads. You can use them just about anywhere you'd normally use a pickle or relish. They're light and tangy with a hint of sweetness and ready the same day you make them.

In a colander, toss the carrots and daikon with the salt and let drain in the sink for about 30 minutes. Give a toss once or twice during this period to shake out as much liquid as possible. As the salt pulls moisture out of the vegetables, they'll become more tender and pliable.

Meanwhile, combine the vinegar, water, and sugar in a small saucepan over medium heat. Stir until the sugar is dissolved, then remove the brine from the heat, and let it cool to room temperature.

Rinse the carrots and daikon under cold running water to remove excess salt, then pack them into jars. Pour the brine over them and let stand for at least 4 hours at room temperature, out of direct sunlight. If you have the time, the pickles are best refrigerated overnight to develop the smooth and sour flavor of traditional *đồ chua*.

CARROT, LEEK, BEEF, AND BARLEY SOUP

MAKES 6 TO 8 SERVINGS

1 tablespoon (15 ml) olive oil

2½ pounds (1.1 kg) beef short ribs

2 leeks (white stems and dark green leaves), sliced (see Preparation Tip on page 139)

4 garlic cloves, smashed with the flat side of a knife

½ cup (120 ml) dry red wine

8 cups (1.9 L) beef broth (see Cook's Note)

2 thyme sprigs

2 bay leaves

3 carrots with greens, carrots diced and greens reserved

1 turnip, diced

½ cup (100 g) pearl barley

1 cup (145 g) shelled peas

Freshly cracked black pepper to taste

Dollop of sour cream for garnishing

Ahhh. This is the type of meal that warms your bones. It's pure comfort in a pot. And if I start to see some clouds move in outside, I'll start cooking this soup in the morning to eat through the rest of the dreary, drizzly day. I usually take some liberty with the beef and barley theme—adding different vegetables here and there, depending on what I have on hand—but three things always remain the same: Short ribs. Whole leeks. And red wine. (To cook with *and* to drink!)

Heat a wide, heavy pot over medium-high heat and add the oil. Brown the short ribs on both sides, about 3 minutes per side, until some of the fat is rendered. Transfer the short ribs to a dish and set aside. Scrape up any brown bits in the pot and add the leeks and garlic, stirring to coat. Pour in the wine, bring to a rapid simmer, and stir until most of the liquid is absorbed, 2 to 3 minutes. Add the short ribs, broth, thyme, and bay leaves to the pot and bring to a boil. Reduce the heat, cover, and simmer for 1 hour. Periodically take a peek under the lid to ensure the liquid maintains a gentle simmer and does not boil.

Remove the lid and add the carrots, turnip, and barley. Increase the heat and bring to a rolling simmer. Continue simmering, uncovered, until the vegetables are tender and the barley is cooked, about 1 hour.

recipe continues

COOK'S NOTE
Now if I had all weekend, I'd make my own beef broth by simmering the short ribs overnight, but I don't think that far ahead. Instead, I deglaze the pot with a splash of red and simmer the ribs in a premade broth, just enough so the meat falls off the bones. In a couple of hours, out comes a thick, hearty, intensely savory soup that's lightened up with a little pat of tangy sour cream.

PREPARATION TIP

This recipe uses both ends of the leeks (the white stems as well as the dark green leaves), so look for long, thick leeks with equal proportions of both. To prepare, trim the green leaves from the white stem. Rinse the leaves under running water to remove any grit in the creases and cut into thin strips crosswise.

Slice the white stems in half lengthwise and rinse under running water, fanning out the layers to wash off any trapped dirt. Cut the stems crosswise into thin half moons and add to the green leaves.

continued

Remove the short ribs from the pot and chop the meat into bite-size pieces. Discard the bones and return the meat to the pot. Strip the carrot leaves from the thicker stems and finely chop the leaves. Measure out 1 cup (60 g) (or 2 large handfuls) of chopped carrot leaves and add them to the pot. Stir in the peas and heat through for 10 minutes. Remove and discard bay leaves. Serve with a few turns of freshly cracked black pepper and a dollop of sour cream. If desired, garnish with a sprinkle of chopped carrot leaves. Store the remaining carrot leaves and reserve for another use.

COOK'S NOTE

The key to a rich, aromatic broth is using meaty, bone-in short ribs. If you don't have any on hand, you can also use bone-in chuck, beef shanks, or oxtails. Avoid the prepackaged "stew meat" sold at the grocery store, as the meat is too lean and will turn tough after a long simmer. You want good, fatty meats that will fall off the bone and add body to the broth. If you're not in any rush to get this meal on the table, you can let the meat simmer for 2 hours or more before adding the barley and vegetables. It will only make it better!

Barley has a tendency to absorb all the broth if it sits in the soup too long, so if you're reheating the next day, you'll want to thin out the soup with some more broth or water.

QUICK-PICKLED SWEET 'N' SPICY
RADISH PODS

MAKES 3 CUPS (700 ML)

---

1 cup (235 ml) water

½ cup (120 ml) rice vinegar

½ cup (120 ml) white wine vinegar

½ cup (100 g) sugar

½ tablespoon kosher salt

2 heaping (150 g) cups radish seed pods

Unless you grow your own radish plants at home, you've likely never seen the long, thin seed pods that emerge once they start flowering. All radish plants produce seed pods that are edible; you may have even come across a patch of wild radishes on a hike and not realized what the pointy-tailed pods were. They appear en masse (a single plant is quite prolific, producing dozens of pods on its flower stalk) at the end of the life cycle, long after the radish root has turned woody and inedible. But even though you can't harvest the radish, consider the seed pods a bonus harvest—a final hurrah—before you pull the plant up.

Pick them while they're green and fresh (because in a few weeks, they'll dry out and turn into actual seeds). The crunchy, peppery pods taste like a concentrated radish and can be easily snacked on by the handful.

Though I like them when they're raw, I love them even more when they're pickled. The sweet and tangy brine mellows out their spice somewhat and turns the radish pods into a zesty topping for sandwiches, tacos, Pea Shoot Salad with Radish and Carrot (page 84), Bottom-of-the-Box Bibimbap (page 180), or even a Bloody Mary.

In a small saucepan over medium heat, combine the water, rice vinegar, wine vinegar, sugar, and salt and stir until the grains are dissolved. Let the brine cool to room temperature.

Pack the radish pods into jars and pour the brine over them, making sure the pods are fully submerged. Pickle at room temperature, out of direct sunlight, for at least 4 hours before serving. For best flavor, pickle overnight in the fridge.

PORTUGUESE SAUSAGE AND RADISH GREEN SOUP

MAKES 6 TO 8 SERVINGS

for the tomato base

1 can (6 ounces, or 170 g) of tomato paste

⅓ cup (80 ml) white vinegar

⅓ cup (115 g) honey

½ teaspoon kosher salt

¼ teaspoon onion powder

¼ teaspoon garlic powder

⅛ teaspoon ground clove

for the soup

1 tablespoon (15 ml) olive oil

1 yellow onion, chopped

4 garlic cloves, minced

10 ounces (280 g) *linguiça* (Portuguese) sausage, thinly sliced on the diagonal

10 ounces (280 g) russet potatoes, cut into small dice

1 can (15 ounces, or 425 g) of cannellini beans, drained and rinsed

8 cups (1.9 L) chicken broth

½ teaspoon ground black pepper

4 cups (152 g) chopped radish greens

¼ cup (60 ml) white vinegar

This tangy soup is a home-style favorite on a brisk fall day. I often serve it with cornbread to sop up the broth, but it's filling enough to be a meal on its own. It was originally inspired by *caldo verde*, a traditional Portuguese soup, but it uses a zesty tomato base that can double as a quick-and-dirty homemade ketchup (should you want to try it by itself).

Mature radish greens from large winter radishes are ideal in this recipe, as they'll wilt beautifully in the broth but still have some texture. For variation, you can also try the soup with beet greens, cabbage, kale, or chard.

To make the tomato base, combine all of its ingredients in a small saucepan over medium heat. Whisk until smooth. Bring to a gentle boil, then reduce the heat and simmer uncovered until the mixture thickens to the consistency of ketchup, about 15 minutes. Stir frequently to keep the mixture from splattering. Remove from the heat and set aside.

Heat a large pot over medium-high heat. Add the oil, onion, and garlic and cook until the onion starts to turn translucent, 2 to 3 minutes. Add the tomato base and stir to coat the onion evenly. Add the sausage, potatoes, beans, broth, and pepper and bring to a boil. Reduce the heat and simmer uncovered, stirring occasionally until the potatoes are tender, about 20 minutes. Add the radish greens and vinegar and continue simmering for 20 minutes more.

SAUSAGE SUBSTITUTIONS

Linguiça is a Portuguese smoked sausage that can often be found in well-stocked supermarkets with other smoked sausages. If it's not available where you are, Portuguese *chouriço* can be used in place of it, as well as the more commonly found Spanish chorizo. In a pinch, kielbasa or another garlicky, spicy smoked sausage will also work here, but the soup will have a slightly different flavor.

BUTTER-BRAISED RADISHES AND RADISH GREENS WITH FARRO

MAKES 4 SERVINGS

---

1 cup (208 g) uncooked whole-grain farro, rinsed

1 pound (455 g) radishes with greens

2 tablespoons (28 g) butter

¾ cup (175 ml) chicken broth

3 tablespoons (45 ml) balsamic vinegar

½ teaspoon kosher salt

¼ teaspoon ground black pepper

¼ cup (38 g) crumbled feta cheese

2 to 3 tablespoons (8 to 12 g) chopped fresh parsley

Most people eat their radishes raw, but have you ever tried cooking them? As with many roots, heat does beautiful things to transform radishes into a totally different vegetable: smoother, mellower, and slightly sweeter.

Cooked radishes have little of the bitter or spicy kick of their crunchy raw selves. But they don't turn tasteless either—a buttery braise forms precious caramelized bits for sweetness, while a balsamic-spiked broth adds tang to their naturally peppery flavor.

I like the bite of whole-grain farro with tender radishes and creamy feta, but feel free to use your favorite grain here.

Bring a large saucepan of salted water to a boil. Add the farro and simmer uncovered for about 30 minutes until the farro is tender but still pleasantly chewy. Drain the farro, cover, and keep warm.

Meanwhile, trim the greens from the radishes and coarsely chop the greens. Cut the radishes into quarters or eighths (depending on how large they are) and set the radishes and greens aside separately.

In a large sauté pan over medium-high heat, melt the butter. Add the radishes in a single layer, with most of the cut sides down, and cook undisturbed for 4 to 5 minutes until the bottoms start to brown and caramelize. Give a stir, then add the broth, vinegar, salt, and pepper, and bring to a boil. Reduce the heat, cover, and simmer for about 10 minutes until the radishes are tender.

Stir in the radish greens and cook for about 3 minutes until wilted. Turn off the heat, add the farro, and toss to combine. Serve warm with a sprinkle of feta and parsley on top.

OVEN-BAKED
POTATO PARMESAN FRIES

MAKES 4 TO 6 SERVINGS

2½ pounds (1.1 kg) russet potatoes, cut into 3-inch (7.5 cm) wedges

2 tablespoons (28 ml) olive oil, plus more for greasing

½ teaspoon kosher salt

½ teaspoon garlic powder

¼ teaspoon smoked paprika

¼ teaspoon ground cumin

¼ teaspoon ground cayenne pepper

¼ cup (25 g) grated Parmesan cheese

2 tablespoons (8 g) chopped fresh parsley

Homemade fries are the best. They're less greasy and more flavorful than fast-food fries and so easy to throw together when you're craving a savory snack. Amp them up with a blend of salty and smoky spices, and you don't even need ketchup—which is how I gauge the perfect fry. (But if you find yourself just waiting around while they bake, you can start a batch of homemade ketchup on the stove using the tomato base recipe from my Portuguese Sausage and Radish Green Soup, page 143.)

Preheat the oven to 450°F (230°C, or gas mark 8). Grease a large rimmed baking sheet and set aside.

In a large bowl, toss the potatoes with the oil until thoroughly coated.

In a small bowl, combine the salt, garlic powder, paprika, cumin, and cayenne pepper. Sprinkle the spices over the potatoes and toss to coat. Spread the potatoes across the prepared baking sheet in a single layer and bake for 30 to 35 minutes until browned and crispy.

Transfer the potatoes to a large serving bowl. While the potatoes are hot, scatter the Parmesan and parsley on top and toss to combine.

SMASHED RED POTATOES
HEAPED ON ROASTED PORTOBELLOS

MAKES 4 SERVINGS

---

for the portobellos

¼ cup (60 ml) balsamic vinegar

1 tablespoon (15 g) Dijon mustard

4 garlic cloves, minced

1 shallot, minced

½ teaspoon kosher salt

A few grinds of black pepper

¼ cup (60 ml) olive oil

4 portobello mushrooms, destemmed and stems reserved

for the potatoes

2 pounds (900 g) red potatoes, cut into 1-inch (2.5 cm) chunks

½ cup (120 ml) half-and-half

½ cup (112 g) butter

2 scallions, sliced

4 garlic cloves, crushed

2 teaspoons minced fresh rosemary

When I cook for vegetarian friends, I often substitute a thick, juicy portobello for the meaty component of the dish (like my Grilled Pepper, Peach, and Portobello Stacks, page 53). The denseness of the mushroom and the savory bite makes it a suitable stand-in for meat. I've even converted many a carnivore with hefty portobellos drenched in zippy marinades and tenderized under high heat.

As such, this is a meatless take on the meat-and-potatoes theme. Rich, meaty mushrooms serve as a succulent base for creamy smashed potatoes piled on high. No peeling needed here—the ruby flecks of skin add a pleasing texture to the mash. Serve this dish as a side to share or as a full meal with a salad.

Preheat the oven to 400°F (200°F or gas mark 6).

In a small bowl, make the marinade by combining the vinegar, mustard, garlic, shallot, salt, and pepper. Whisk in the oil until well blended. Place the portobellos on a large rimmed baking sheet and pour the marinade evenly over them. Let stand at room temperature for at least 30 minutes for the flavors to meld together. Roast the portobellos until juicy and tender, about 20 minutes.

Meanwhile, bring a large pot of salted water to a boil and add the potatoes. Boil until a fork easily pierces the flesh, about 10 minutes. Drain and keep warm in the pot.

In a small saucepan over medium heat, combine the half-and-half, butter, scallions, garlic, and rosemary. Finely chop the reserved portobello stems and stir them into the sauce. Gently simmer until the butter is melted. Pour the sauce over the potatoes and mash with a potato masher until all the liquid is incorporated. Be careful not to overmash; the potatoes should have a soft, chunky texture with strips of red skins throughout. Heap the mashed potatoes onto the portobellos and serve warm.

SAVORY SWEET POTATO
HUMMUS

MAKES 4 CUPS (946 ML)

---

1 pound (455 g) sweet potatoes, cut into 1-inch (2.5 cm) cubes

1 can (15 ounces, or 425 g) of chickpeas, drained and rinsed

4 garlic cloves

¼ cup (60 ml) olive oil, plus more for serving

1 teaspoon ground cayenne pepper

1 teaspoon ground cumin

½ teaspoon kosher salt

Zest of 1 lemon

Juice of ½ lemon

Smoked paprika for serving

While sweet potatoes are generally associated with sweet flavors, this savory and spicy hummus will have you thinking about them in new ways. A blend of tender sweet potatoes and smoky spices changes up the usual hummus routine of garlic and tahini (and gives it a brilliant orange color). It makes an addictive dip for flatbread, pita chips, carrot sticks, celery nubs, and the list goes on. I'll sometimes smear it onto a Mediterranean-style wrap with roasted vegetables; the creaminess eliminates the need for a messy sauce.

Bring a large saucepan of water to a boil. Add the sweet potatoes and boil until very soft, about 10 minutes. Drain, reserving about ½ cup (120 ml) of the water, and let cool.

Add the remaining ingredients (except the smoked paprika) to a food processor and pulse continuously until smooth. If you're sensitive to spice, I suggest starting with only ½ teaspoon cayenne pepper and adding more to taste. As your mixture transforms into creamy hummus, you can add some of the reserved sweet potato water (1 tablespoon [15 ml] at a time) if you prefer a looser consistency. Serve with a generous swirl of oil on top and a dusting of smoked paprika.

COOK'S NOTE

I use orange-fleshed sweet potatoes as they tend to be softer and smoother than the white-fleshed variety when cooked, and they blend beautifully with creamy chickpeas.

If you have the time and inclination, soaking and cooking your own chickpeas will take this hummus to another level. You'll need about 2 cups (328 g) of cooked chickpeas (from about ⅔ cup [133 g] dried chickpeas) for this recipe. Of course, there's nothing wrong with using the canned chickpeas recommended here. They're readily available and store easily in the pantry for those spur-of-the-moment hummus cravings.

CREAMY SWEET POTATO SOUP
WITH MAPLE SYRUP

MAKES 6 TO 8 SERVINGS

¼ cup (55 g) butter

1 yellow onion, chopped

4 garlic cloves, chopped

3 pounds (1.4 kg) sweet potatoes, cut into 1-inch (2.5 cm) cubes

4 thyme sprigs

6 cups (1.4 L) chicken broth

½ cup (115 g) plain yogurt

3 to 4 tablespoons (45 to 60 ml) maple syrup

Dollop of yogurt for garnishing

Toasted and chopped pecans for garnishing

Crumbled blue cheese for garnishing

Days getting shorter, leaves turning colors, and the arrival of freshly harvested sweet potatoes—these are all signs of the start of a new season.

As soon as those first tubers appear at the farmers' market, I know that fall is on its way. Those first chilly whips in the air mean I'm never far from the stove, stirring up a pot of this warm, creamy soup and tucking into a bowl for the night. The maple syrup plays on the sweetness of sweet potatoes but doesn't turn the dish too sweet; of course, you can add another spoonful of syrup if you're so inclined.

In a large pot over medium-high heat, melt the butter. Stir in the onions and garlic and cook for 2 to 3 minutes until the onions start to turn translucent. Add the sweet potatoes, thyme, and broth and bring to a boil. Reduce the heat and simmer uncovered until the sweet potatoes are very soft, 20 to 25 minutes.

Discard the thyme. Using an immersion blender, puree the soup until smooth. Stir in the yogurt and maple syrup (using more or less maple syrup to your taste preference) and heat through. Serve with a dollop of yogurt, a small handful of pecans, and a spoonful of blue cheese on top, if desired.

SPICY STIR-FRIED
SWEET POTATO LEAVES

MAKES 4 SERVINGS

for the sauce

¼ cup (60 ml) water

2 tablespoons (28 ml) fish sauce

1 tablespoon (13 g) sugar

1 garlic clove, minced

½ serrano pepper, minced

Juice of ½ lime

for the stir-fry

2 tablespoons (28 ml) sunflower oil

1¼ pounds (570 g) sweet potato leaves, coarsely chopped

4 garlic cloves, minced

½ serrano pepper, minced (optional, if you want more heat)

Sweet potato leaves are most abundant in summer, when the lush vines find their way to Asian markets and farmers' markets before their tuberous counterparts appear in fall. They're especially popular in Asian cooking, where the mild and tender leaves are used like any other tender green in soups, sautés, and stir-fries.

One of my favorite recipes is this classic Vietnamese version that quickly wilts the leaves and douses them in a tangy, traditional fish sauce called *nước chấm*. This pungent sauce is a mainstay of Vietnamese cuisine and ticks off all the boxes when it comes to artfully blending sweet, sour, salty, and spicy.

In a small bowl, combine all the ingredients for the sauce. Stir until the sugar is dissolved and set aside.

Heat a wok or large sauté pan over high heat. Add the oil and a few large handfuls of sweet potato leaves. Quickly toss them around until they start to wilt and reduce in volume, about 1 minute. Add the garlic, serrano (if using), and another few handfuls of sweet potato leaves and continue stir-frying for 2 to 3 minutes until all the leaves are wilted and the stems are crisp-tender. Pour in the sauce and stir to coat. Remove from the heat and serve immediately.

MYTH BUSTING

Contrary to popular belief, sweet potatoes are not related to potatoes. They belong to the morning glory family (*Convolvulaceae*), which is most apparent when their morning glory-like flowers start blooming. Sweet potatoes are true root-to-leaf vegetables (sort of like the veggie counterpart to nose-to-tail), so both ends are edible. If you grow your own crop at home, harvest the young leaves and shoots (as you'll have plenty of them all summer) before you dig up the tubers in fall.

SAIGON CURRY WITH
SWEET POTATOES AND CARROTS

MAKES 4 TO 6 SERVINGS

2 pounds (900 g) skinless bone-in chicken thighs

2½ tablespoons (16 g) Madras curry powder (or Vietnamese curry blend), divided

1½ teaspoons kosher salt, divided

½ teaspoon ground black pepper, divided

1 tablespoon (15 ml) sunflower oil

3 lemongrass stalks, cut into 3-inch (7.5 cm) segments and bruised with the flat side of a knife

1-inch (2.5 cm) piece ginger, minced

4 garlic cloves, minced

½ teaspoon red pepper flakes

4 cups (946 ml) chicken broth

1 can (13.5 ounces, or 380 g) of coconut milk

1 pound (455 g) sweet potatoes, cut into 1-inch (2.5 cm) chunks

1 pound (455 g) carrots, sliced in half lengthwise and cut into 2-inch (5 cm) segments

1 yellow onion, cut lengthwise into eighths

Chopped cilantro for garnishing

Sliced scallions for garnishing

This classic Vietnamese comfort dish, *cà ri gà* (chicken curry), is truly a melting pot of flavors. The fragrant curry comes from Indian spices, the fiery red pepper and sweet coconut are inspired by Thai cooking, and the crusty baguette used to sop up all the curry is a nod to Vietnam's French colonial influence.

Growing up, a giant pot of *cà ri gà* simmering on the stove was to me the way a homey beef stew was to my friends. This is Vietnamese soul food at its finest: full of aromatics, easy to prepare, and everyone has her own way of making it. In my version, I use sweet potatoes (in place of the more traditional potatoes) to bring out the sweetness of the curry. I like to serve it with a freshly baked baguette, torn by hand and dipped into the stew, but it also makes a hearty meal over rice.

In a shallow dish, dust the chicken with ½ tablespoon curry powder, ¾ teaspoon salt, and ¼ teaspoon black pepper. Heat a wide, heavy pot over medium-high heat and add the oil. Arrange the chicken in a single layer in the pot and sear until golden brown on the bottom, about 3 minutes. Flip the chicken and sear the other side, about 3 minutes. Transfer the chicken to a clean dish and set aside.

Let the oil and juices reheat in the pot and add the lemongrass, ginger, garlic, and red pepper flakes. Cook until the herbs and spices are fragrant, 2 to 3 minutes. While they cook, cut the chicken into bite-size chunks but leave the larger pieces on the bones to fall off naturally as they simmer. Add the chicken chunks and bones to the pot and pour in the broth and coconut milk. Stir to blend and add the remaining 2 tablespoons (13 g) curry powder, ¾ teaspoon salt, and ¼ teaspoon black pepper. Bring to a boil.

Reduce the heat and simmer uncovered for 20 minutes. Add the sweet potatoes, carrots, and onion and bring back up to a boil. Once the liquid reaches a continuous boil, reduce the heat and simmer until the chicken is tender and the vegetables are soft but not mushy, about 20 minutes. Discard the lemongrass before serving and garnish with cilantro and scallions, if desired.

7

MELONS & GOURDS

Melons and gourds all belong to the *Cucurbitaceae* family, which includes watermelons, cucumbers, summer squash, and winter squash. It's no wonder, then, that they share some scrumptious similarities on the taste buds. Watermelon seeds and pumpkin seeds take on a nutty crunch when toasted, watermelon rind tastes like a cross of cucumber and zucchini, and cooked cucumber can easily pass for pattypan squash in its sweetness and juiciness.

Most of us are familiar with cucumber as a fresh, raw ingredient for salads and tzatziki, but cooking brings out a completely new side of it. Like many other gourds, cucumbers have edible fruits, flowers, and leaves, though we typically see only the fruits sold. The tender sprouts and shoots of cucumber plants are similar in flavor to cucumber skin, meaning they can sometimes be bitter, depending on the variety. (But if you like bitter, by all means, try a nibble.)

When it comes to most bang for the buck, though, summer squash and winter squash give it their all. Every part of the plant is not only edible but downright delicious. The vines, flowers, flesh, and seeds of squash, whether it's your garden-variety zucchini or a French heirloom pumpkin, range in flavor from light and sweet to deep and nutty.

If you find yourself with baby squash with blossoms still attached—or better yet, you grow your own crop at home—pick off each of these parts to sample all the different textures and subtleties in flavor.

WATERMELON RIND
AND JALAPEÑO PICKLES

MAKES 4 CUPS (946 ML)

1 cup (235 ml) cider vinegar

1 cup (235 ml) water

¼ cup (50 g) sugar

3 tablespoons (42 g) kosher salt

3½ cups (525 g) sliced watermelon rind (see Preparation Tip, page 161)

Zest of 1 lime

1 jalapeño pepper, thinly sliced (remove ribs and seeds for less heat)

I've always found it strange that we devour a watermelon right up to the white rind, but leave the rest of it uneaten. The rind might not be as sweet as the flesh, but it's reminiscent of a melon crossed with cucumber: fresh, crisp, and mild. In fact, you can pickle it much like you'd pickle a cucumber and eat it the same way.

I like these pickles piled high on a meaty, greasy taco (I admit I do have an occasional weakness for grease), where the sweet, sour, and spicy notes do a dance party on the tongue.

In a small saucepan over medium-high heat, bring the vinegar, water, sugar, and salt to a simmer. Stir until the sugar is dissolved, then turn off the heat and let cool to room temperature.

Meanwhile, toss the watermelon rind with the lime zest. Layer the watermelon rind and jalapeño in two pint (473 ml) jars (or one quart jar [946 ml]) and then pour the brine over them. Let stand at room temperature, out of direct sunlight, for at least 4 hours before serving. For best flavor, refrigerate the pickles overnight to let the flavors develop more deeply.

COOK'S NOTE
A baby watermelon weighing around 4½ pounds (2 kg) should yield approximately 3 cups (450 g) of watermelon rind. While you can slice the watermelon rind however you'd like, I prefer to peel paper-thin wide "ribbons" of rind about 3 inches (7.5 cm) long for these pickles.

PREPARATION TIP

The easiest kind of watermelon rind to prepare (and the best kind to eat) is about ¾ to 1 inch (2 to 2.5 cm) thick. How can you tell whether a watermelon has a thick rind? Just ask the farmer or vendor or cut one open and make note of that particular variety's characteristics.

To start, slice the watermelon in half and place the cut side down. Peel the green skin (a Y-shaped peeler makes easy work of this) and cut the watermelon into thick wedges. Slice off the pink flesh and reserve for another use (like my Watermelon, Tomato, and Basil Salad with Tangy Red Onion, page 162). Repeat with the other half until you're left with only the white rind from the whole watermelon. Peel, slice, or chop the rind as needed for your recipes.

WATERMELON, TOMATO, AND BASIL SALAD
WITH TANGY RED ONION

MAKES 4 SERVINGS

Heaping ½ cup (80 g) thinly sliced red onion

¼ cup (60 ml) red wine vinegar

2 heaping cups (360 g) wedged tomatoes

4 heaping cups (600 g) seeded and cubed watermelon

3 tablespoons (8 g) thinly sliced fresh basil

The name alone might not hook you at first glance, but one taste of this vibrant salad and I swear you'll be swooning. It combines all the boldest flavors of summer into one simple dish that's welcome at every picnic or potluck. The secret is in the red wine vinegar: it adds a bit of tang to brighten the watermelon and soften the bite of the red onion. (I've been known to add even more vinegar to the salad after it's assembled.)

Because this salad is so seasonal, the key is using the freshest ingredients you can find: sweet, juicy watermelon; vine-ripened heirloom tomatoes; and rich, fragrant basil. If there was ever a time to seek out beautifully ribbed and brilliantly colored tomatoes to make a dish stand out, this would be it!

In a small bowl, soak the red onion in vinegar for 30 minutes.

Meanwhile, place the tomatoes in a colander and let drain in the sink until ready to use.

Gently combine the watermelon, tomatoes, and basil in a large serving bowl. Lightly toss with the onion and vinegar and refrigerate for a few hours before serving.

STIR-FRIED
WATERMELON RIND

MAKES 6 SERVINGS

for the stir-fry

4 cups (600 g) sliced watermelon rind (see Preparation Tip on page 161)

1 teaspoon kosher salt

2 tablespoons (28 ml) sunflower oil

4 garlic cloves, minced

1-inch (2.5 cm) piece ginger, minced

2 cups (208 g) bean sprouts

1 carrot, julienned or cut into 3-inch (7.5 cm) matchsticks

4 scallions, sliced on the diagonal into 3-inch (7.5 cm) segments

Freshly cracked black pepper to taste

for the sauce

2 tablespoons (28 ml) soy sauce

1 tablespoon (15 ml) rice vinegar

1 tablespoon (15 ml) sesame oil

Most people have heard of pickling the watermelon rind, but cooking it? Like the Pan-Fried Cucumber in Honey Sesame Sauce (page 169), this part of the fruit is typically used raw, if it's used at all. But take a bite of the crunchy white rind after you've eaten the rest of the watermelon, and you'll find that it tastes a little like cucumber and a little like melon. Toss it around over a high flame and its texture turns tender and zucchini-like. It's refreshing with a hint of sweetness, and it doesn't turn this savory stir-fry into a fruit salad.

If you scale up the sauce, this dish makes a great topping for an Asian noodle bowl.

Toss the watermelon rind with the salt in a colander, then let it drain in the sink while you prepare the remaining ingredients, at least 15 minutes.

In a small bowl, combine all the ingredients for the sauce and set aside.

Heat a wok or large skillet over medium-high heat. Meanwhile, shake up the watermelon rind with your hands and squeeze out any excess liquid. Return to the wok and add the oil, garlic, and ginger and stir-fry until fragrant, about 30 seconds. Add the watermelon rind and stir-fry for 2 minutes. Add the bean sprouts, carrot, and scallions and toss them around until the carrot is tender and the scallions are wilted, 2 to 3 minutes. Pour in the sauce and stir to coat. Serve with a few turns of freshly cracked black pepper on top.

COOK'S NOTE
In this recipe, I slice the watermelon rind into matchsticks for uniformity with the other vegetables.

QUICK–
PICKLED CUCUMBERS

Quick-pickled cucumbers make a crisp, piquant, and delectable side to many savory dishes. I like to pair them with grilled meats, but they are also an easy, satisfying snack right out of the jar. Sweet Japanese pickles are typically sliced paper-thin with a mandoline, while spicy Korean pickles are usually sliced 1/8 to 1/4 inch (3 to 6 mm) thick. Both of these recipes call for Japanese cucumbers, but you can substitute any mild, thin-skinned, slender, and minimally seeded variety, such as English or Persian cucumbers.

Spicy Korean Quick-Pickled Cucumbers (top);
Sweet Japanese Quick-Pickled Cucumbers (bottom)

SWEET JAPANESE QUICK-PICKLED CUCUMBERS

MAKES 2 CUPS (473 ML)

1 pound (455 g) Japanese cucumbers, thinly sliced

2 teaspoons kosher salt

¼ cup (60 ml) rice vinegar

1½ tablespoons (20 g) sugar

1 tablespoon (8 g) toasted sesame seeds

In a colander, toss the cucumbers with the salt and let drain in the sink for 1 hour. Give them a good toss once or twice to shake off as much liquid as possible.

In a medium bowl, combine the vinegar, sugar, and sesame seeds. Add the cucumbers and toss to coat. Cover and chill before serving. The pickles can be served the same day, but are smoothest in flavor in 2 to 3 days.

SPICY KOREAN QUICK-PICKLED CUCUMBERS

MAKES 2 CUPS (473 ML)

1 pound (455 g) Japanese cucumbers, thinly sliced

2 teaspoons kosher salt

2 tablespoons (28 ml) rice vinegar

1 tablespoon (15 ml) sesame oil

1 tablespoon (15 ml) sesame seeds

2 teaspoons (8 g) gochugaru (Korean red pepper powder, see sidebar)

1 teaspoon sugar

1 scallion, thinly sliced

1 garlic clove, minced

In a colander, toss the cucumbers with the salt and let drain in the sink for 1 hour. Give them a good toss once or twice to shake off as much liquid as possible.

In a large bowl, combine the vinegar, oil, sesame seeds, gochugaru, sugar, scallion, and garlic. Add the cucumbers and toss to coat. Cover and chill before serving.

NOT ALL PEPPER POWDERS ARE ALIKE

If you've never tried kimchi, think of it as the Korean version of sauerkraut—spicy and pungent and packed with all the probiotic goodness of fermentation. It all starts with gochugaru, a Korean red pepper powder (sometimes also called Korean chile flakes or Korean chile powder) that brings smoke and fire to the rich and complex blend of flavors. Gochugaru can be found in every Korean market as well as online; do not confuse it (or replace it) with American red pepper flakes or American chili powder, as the spices are not interchangeable in these recipes.

CALIFORNIA BREAKFAST BOWL WITH CUCUMBER SPROUTS

MAKES 4 SERVINGS

MAKES 4 SERVINGS

for the quinoa

1 tablespoon (15 ml) olive oil

2 cups (346 g) quinoa, rinsed

3 cups (700 ml) water

½ teaspoon kosher salt

for the breakfast bowl

2 tablespoons (28 ml) olive oil, divided

1 can (15 ounces, or 425 g) of black beans, drained and rinsed

4 eggs, beaten with a pinch of kosher salt

1 avocado, thinly sliced or diced

2 cups (240 g) packed cucumber sprouts

1 cup (260 g) salsa

½ cup (75 g) crumbled Cotija cheese

COOK'S NOTE
If you can't source cucumber sprouts, top your bowls with sunflower sprouts, daikon sprouts, broccoli sprouts, pea sprouts, or any other sprout or microgreen.

Alfalfa sprouts, bean sprouts, pea sprouts, sunflower sprouts, sprouted grains, and all manner of microgreens—inside all those little leaves are a powerhouse of phytonutrients, even moreso than their mature counterparts. In fact, if you grow your own herbs and vegetables and ever had to thin out a row of seedlings, you inadvertently found yourself with a handful of nutritious sprouts for a salad.

Cucumber sprouts were one of those happy accidents in my garden. Every other sprout from a row of thickly seeded cucumber plants would normally go in the compost pile, but a curious bite one afternoon yielded a pleasant surprise. Cucumber sprouts have a light, refreshing crunch with the bold flavor of cucumber skin. The sweeter the variety, the sweeter the sprouts. Tossed into a bowl of quinoa with California avocado, black beans, scrambled eggs, salsa, and Cotija, it's the kind of breakfast I inhale after a morning surf when I'm starving and want something filling, fresh, and homemade.

To prepare the quinoa, heat a large saucepan over medium-high heat and drizzle in the oil. Add the quinoa and toast for 3 to 4 minutes, stirring occasionally, until it releases a nutty fragrance. Add the water and salt and bring to a boil. Reduce the heat, cover, and simmer until most of the liquid is absorbed, about 15 minutes.

Meanwhile, heat a medium skillet over medium heat. Add 1 tablespoon (15 ml) of the oil and the beans and heat through until warm, 3 to 5 minutes. Transfer the beans to a dish and keep warm. Wipe out the skillet and reheat over medium heat. Add the remaining 1 tablespoon (15 ml) of oil and pour in the eggs to coat the bottom. Gently and constantly stir the eggs around until they are barely set. (I like to use chopsticks for softly scrambling eggs, as they're better for breaking them down into smaller curds.) Don't overcook the eggs; they should still look wet, but not runny.

When the quinoa is done, turn off the heat and let it stand, covered, for about 5 minutes. Fluff up the quinoa with a fork and divide among four bowls. Top each bowl with an equal helping of beans, eggs, avocado, cucumber sprouts, salsa, and Cotija and serve warm.

SPROUTING ANYTIME, ANYWHERE

While the cucumber plant is typically grown for the fruit, you can sow a flat of cucumber seeds indoors, year round, just for the sprouts. In warm temperatures, the sprouts appear within a week. Harvest when the leaves are small and the stems are still tender.

PAN-FRIED CUCUMBER
IN HONEY SESAME SAUCE

for the cucumbers

2 pounds (900 g) cucumbers

1 teaspoon kosher salt

2 tablespoons (28 ml) sunflower oil

4 garlic cloves, minced

1 serrano pepper, minced (remove ribs and seeds for less heat)

for the sauce

1 tablespoon (15 ml) soy sauce

1 tablespoon (15 ml) rice vinegar

1 tablespoon (15 ml) sesame oil

1 tablespoon (20 g) honey

1 teaspoon toasted sesame seeds

Don't let the thought of cooked cucumber turn you off from this dish. When it meets the heat of a frying pan, cucumber turns tender and juicy much like zucchini does and tastes as mild as any summer squash (with which it belongs in the same family, *Cucurbitaceae*, or the gourd family).

This type of dish is quite common in Chinese cuisine; in fact, many traditionally raised Chinese kids probably ate a cooked cucumber before they ever ate a raw or pickled cucumber. (If it weren't for my dual upbringing with a Vietnamese mother, whose culture embraced lots of fresh vegetables, I don't think I would've had a raw cucumber until I started eating in my grade-school cafeteria!) For Westerners, it's an unexpected way to eat an ordinary cucumber, and pan-frying brings out a whole new side of it that everyone should try at least once.

To prepare the cucumbers, slice each one in half lengthwise. Run a small spoon down the center to scoop out the seeds and cut the cucumber crosswise into ¼-inch (6 mm) slices. Place all the cucumber slices in a colander, toss with the salt, and let them drain in the sink for 15 minutes.

Meanwhile, make the sauce by combining all of its ingredients in a small bowl and set aside.

Heat a wok or large skillet over high heat and coat the surface with the sunflower oil. Toss the cucumbers with your hands to shake off any excess liquid, then spread them across the wok with as little overlap as possible. Fry undisturbed until the cucumbers begin to brown on the bottom, about 2 minutes. Using a spatula, flip the cucumbers and fry until the other side begins to color, about 2 minutes. Stir in the garlic and serrano and cook until fragrant, about 30 seconds. Pour in the sauce, stir to coat, and heat through.

TOASTED PUMPKIN SEEDS

GARLIC PUMPKIN SEEDS

MAKES 1 CUP (235 ML)

1 cup (64 g) pumpkin seeds

1 tablespoon (15 ml) olive oil

1 teaspoon Worcestershire sauce

1 teaspoon garlic powder

CURRIED PUMPKIN SEEDS

MAKES 1 CUP (235 ML)

1 cup (64 g) pumpkin seeds

1 tablespoon (15 ml) melted coconut oil

2 teaspoons curry powder

SWEET 'N SPICY PUMPKIN SEEDS

MAKES 1 CUP (235 ML)

1 cup (64 g) pumpkin seeds

1 tablespoon (15 ml) olive oil

2 teaspoons packed brown sugar

½ teaspoon ground cinnamon

½ teaspoon ground cayenne pepper

¼ teaspoon ground cumin

All seeds from the squash family (*Cucurbitaceae*) are edible, but we see them most often in the form of pumpkin seeds, since one carving pumpkin will yield at least a cup or two (64 to 128 g) of seeds. They're made more snack-worthy by toasting the shells and giving them crunch, so you can pop the whole seed in your mouth, shell and all.

I like to boil my pumpkin seeds in a simple brine first, which infuses them with a subtle salty flavor and gives them a little extra snap. My favorite—and recommended—way to eat them is by the handful, washed down with a cold pumpkin ale!

PREPARATION FOR ALL RECIPES

Remove as much of the pulp as possible from your pumpkin seeds. (But no need to be fanatical; boiling them will remove the rest of the stringy bits.)

In a small saucepan, bring 3 cups (700 ml) of water and 2 tablespoons (28 g) of kosher salt to a boil (while not necessary, a drizzle of oil helps prevent boil over). Add 1 cup (64 g) of pumpkin seeds and boil for 10 minutes. Drain and shake dry. Then, choose your favorite flavor combination and continue with the recipe.

INSTRUCTIONS FOR ALL RECIPES

Preheat the oven to 375°F (190°C, or gas mark 5). Line a large rimmed baking sheet with parchment paper and set aside.

In a medium bowl, toss all the ingredients together until evenly coated. Spread the seeds across the prepared baking sheet in a single layer and bake for 15 to 20 minutes, stirring halfway through, until the seeds are crisp, dried, and slightly darkened (but not blackened) on all sides. Larger seeds may require longer toasting times and smaller seeds may require shorter toasting times, so be sure to keep an eye on your oven to avoid burning them.

Generally, I know the pumpkin seeds are perfectly toasted when I hear the shells start to pop. The shells should feel light and airy and be easy to bite through; if they're chewy, give them a little more time in the oven. The inner seeds should be tender with a deep green or golden brown hue.

SRIRACHA GINGER PUMPKIN SEEDS

MAKES 1 CUP (235 ML)

1 cup (64 g) pumpkin seeds

1 tablespoon (15 ml) olive oil

1 teaspoon sriracha

¼ teaspoon ground ginger

¼ teaspoon garlic powder

SICILIAN SQUASH SHOOT SOUP

MAKES 6 TO 8 SERVINGS

8 cups (1.9 L) water

1 teaspoon kosher salt

3 cups (180 g) coarsely chopped squash shoots (see Preparation Tip on page 173)

2 new potatoes, cut into small dice

6 ounces (170 g) dried spaghettini (thin spaghetti), snapped into 3-inch (7.5 cm) lengths

¼ cup (60 ml) olive oil, plus more for serving

½ yellow onion, chopped

2 garlic cloves, minced

¼ teaspoon red pepper flakes

1 pound (455 g) plum tomatoes, chopped

¼ cup (6 g) packed fresh basil, sliced

¼ teaspoon ground black pepper

4 cups (946 ml) chicken broth

Grated Parmesan cheese for serving

If you ask most kids to name their everyday vegetables, they'll likely answer broccoli, lettuce, or spinach. If you'd asked me as a kid to name the everyday vegetables in my house, squash shoots would have made that same list. Strange as it sounds, I never actually thought about where they came from until I started growing my own squash crop a few years ago. Seeing those long, vigorous vines sprawl and snake their way across my garden beds was almost an epiphany—so *that's* what I was eating all those years.

Squash shoots have a delicate green flavor with hints of melon. While they're a staple in Southeast Asian cooking (usually sautéed in garlic and chiles or stewed in a coconut curry), they're also common in Sicilian cooking, where the stems, leaves, and tendrils are called *tenerumi*. Both cultures use the tender tips of the opo squash (what you might also know as long melon, bottle gourd, calabash, or *cucuzza*, if you're Italian), though the shoots from any squash plant are edible. In the summer when these greens are abundant, throw them into a rustic vegetable soup the way the Sicilians do—in a classic *minestra di tenerumi* that marries the season's freshest flavors.

Combine the water and salt in a large pot and bring to a boil. Add the squash shoots and potatoes and cook until tender, about 10 minutes. Using a slotted spoon, transfer the vegetables to a dish and set aside.

Bring the pot of water back up to a boil and add the spaghettini. Cook according to the package directions until al dente.

Meanwhile, heat a medium saucepan over medium-high heat. Add the oil, onion, garlic, and red pepper flakes and cook until the onion starts to turn translucent, 2 to 3 minutes. Add the tomatoes and cook, stirring occasionally until they soften and start to collapse, 3 to 5 minutes.

Stir the tomato sauce into the pot of spaghettini. Add the basil, black pepper, squash shoots, potatoes, and broth and heat through. Serve with a handful of grated Parmesan on top.

PREPARATION TIP

If your squash shoots came from a farm or an Asian market, they'll likely be bundled together in long stems. Before using squash shoots, you'll want to trim them so they don't become a chewy, stringy mess in the soup. Snip off all the tendrils and remove the tough, prickly, lower parts of the stem. The easiest way to determine whether a stem is usable or not is to bend it in several places along its length; any part that does not bend easily should be trimmed. If you're using squash shoots from your own plants, harvest only the tender tips, leaves, and stems from the last few inches (7.5 to 10 cm) of the vine.

GINGERED
BUTTERNUT BISQUE

MAKES 6 TO 8 SERVINGS

1 tablespoon (15 ml) olive oil

1 yellow onion, chopped

2-inch (5 cm) piece ginger, chopped

4 garlic cloves, chopped

6 cups (840 g) diced butternut squash

3 thyme sprigs

6 cups (1.4 L) chicken broth

1/2 teaspoon ground cinnamon

1/2 teaspoon ground nutmeg

1/2 teaspoon ground black pepper

1 cup (235 ml) heavy cream, plus more for garnishing

Chopped fresh thyme for garnishing

In this rich and creamy bisque, sweet butternut squash is balanced with spicy ginger and a touch of cinnamon and nutmeg, a combination that fills the air with the most intoxicating scent as it's simmering on your stove. It's almost like one of those scented candles you can buy in a jar, only you'd be hard-pressed to find the same home-made, sense-tingling, warms-you-to-your-core scent that lingers long after you've extinguished the flame.

This is a soup that feels familiar and comforting, the kind of soup that makes you want to bundle up in a sweater and cozy up next to a fire. Serve it as a meal in itself with a loaf of crusty bread and a glass of wine.

Heat a large pot over medium-high heat. Drizzle in the oil and add the onion, ginger, and garlic. Cook until the onion starts to turn translucent, 2 to 3 minutes. Add the squash, thyme, and broth and bring to a boil. Stir in the cinnamon, nutmeg, and pepper, then reduce the heat and simmer uncovered, stirring occasionally until the squash is tender, about 20 minutes.

Discard the thyme. Using an immersion blender, puree the soup until smooth. Stir in the heavy cream and heat through. Serve with a swirl of heavy cream and a sprinkle of chopped fresh thyme on top, if desired.

SQUASH BLOSSOM AND
ROASTED POBLANO TACOS

4 poblano chiles

½ pound (225 g) squash blossoms with baby squash

1 tablespoon (15 ml) olive oil

½ white onion, finely chopped

2 garlic cloves, minced

A few pinches of kosher salt

A few grinds of black pepper

8 corn tortillas, warmed (if your tortillas are very thin, double them up for the tacos)

Crumbled Cotija cheese for serving

Crema Mexicana (Mexican cream) or sour cream for serving

Of all the flowers you can eat, squash blossoms are among my favorite. Not only are they big, bright, and beautiful, they happen to be palatable, too (unlike other edible flowers, which I feel are best left on the plant). Tasting like a cross of zucchini and cucumber with a velvety texture, squash blossoms are a summer delicacy. They're often found in abundance as standalone flowers, since male blossoms (which show up first and grow off a long, thin stem) are more commonly harvested than female blossoms (which bear the fruit). But I think it's a real treat when you can find female blossoms with baby squash attached; you get two different tastes and textures from the same plant.

Mexican cooking makes especially good use of squash blossoms, where they're called *flores de calabaza*. They're often stuffed with cheese and fried in oil, but I like their delicate flavor best in a simple sauté with onion and garlic. My preference is zucchini blossoms, as they tend to be larger with larger fruit, but any baby squash with blossoms will work for this recipe.

Set the poblanos over an open flame on a stove burner and char on all sides, turning them with metal tongs, until the skins are blistered and blackened. (Alternatively, preheat the broiler. Place the poblanos in a broiler pan and position it 2 to 3 inches (5 to 7.5 cm) from the heating element. Char the poblanos on all sides, turning once or twice until the skins are blackened, 5 to 10 minutes.) Place the poblanos in a plastic bag (or a bowl fitted with a lid), seal, and let them steam for 15 to 20 minutes.

Meanwhile, prepare the squash blossoms. Detach the blossoms from the squash and rinse under running water to wash away any bugs that might be lingering inside. Thoroughly pat dry with a towel. (Some people remove the pistils from the centers, but they're completely edible and have a pleasant texture akin to button mushrooms; I always keep mine intact.) Slice the blossoms in half lengthwise and set aside. Thinly slice the squash and set them aside separately from the blossoms.

Remove the poblanos from the plastic bag and trim the stems. Discard the seeds, peel the skins (they should lift off easily with your fingers), and slice the poblanos into ¼-inch (6 mm)-wide strips. Set aside.

Heat a large skillet over medium-high heat and drizzle in the oil. Add the onion and garlic and cook until the onion starts to turn translucent, 2 to 3 minutes. Add the squash and cook until tender, about 3 minutes. Add the squash blossoms and sprinkle the salt and pepper on top. Cook until the blossoms are wilted, about 3 minutes.

To assemble the tacos, fill each tortilla with an equal heaping of the poblanos and the squash mixture. Serve with a sprinkle of crumbled Cotija and a drizzle of crema Mexicana.

COOK'S NOTE
Squash blossoms appear at farmers' markets and Mexican markets all summer. If you can't find female blossoms with baby squash attached, use about 1 heaping cup (235 ml) each of male blossoms (halved lengthwise) and tender squash (finely chopped).

ZUCCHINI NOODLES
WITH ROASTED TOMATOES, PESTO, AND PINE NUTS

MAKES 4 SERVINGS

1¼ pounds (570 g) cherry tomatoes, halved

3 garlic cloves, sliced

3 tablespoons (45 ml) olive oil

A few pinches of kosher salt

A few grinds of black pepper

3 rosemary sprigs

3 thyme sprigs

2 pounds (900 g) zucchini, julienned

½ cup (130 g) pesto (homemade or store-bought)

¼ cup (35 g) toasted pine nuts

Shaved Parmesan cheese for serving

These zucchini noodles (zoodles?) are like pasta in disguise, and that's what I love about them. They're every bit as fun to twirl around a fork and they play well with pesto, but they leave you with none of that heavy, bloated feeling after a heaping plateful. Since the zucchini brings its own subtle flavor to the table, no heavy sauce is needed for this "pasta" dish either. For a bona fide noodle fiend like myself, I can tell you that I almost don't miss real noodles after this.

Try it with my Tomato Leaf Pesto (page 35) for a more classic pesto flavor or gather up your odds and ends to make my Kale Stem Pesto (page 57), which adds a bright, lemony note to this dish.

Preheat the oven to 325°F (170°C, or gas mark 3).

In a large bowl, toss the tomatoes with the garlic, oil, salt, and pepper until evenly coated. Spread the tomatoes across a large rimmed baking sheet in a single layer, cut-sides up, and scatter the rosemary and thyme on top. Roast for about 1 hour until the tomatoes are softened and shriveled. Discard the dried herbs.

In a large sauté pan over medium-high heat, combine the zucchini and pesto and heat through for about 2 minutes.

To serve, divide the zucchini among four plates. Top each plate with a generous heap of roasted tomatoes (including all the juices), a spoonful of pine nuts, and some shaved Parmesan. Serve warm or at room temperature.

PREPARATION TIP

To get the most out of your zucchini, run the julienne peeler down its length, spin the zucchini a bit, then peel it again. Move around the zucchini in a circle and make even julienne cuts until you reach the soft and seedy core (but don't discard the core, chop it up into a soup or sauté later). Larger zucchini will give you longer, more noodle-like strands, so this is an ideal way to use up any overgrown zucchini from your garden. Now, if you have a monstrous zucchini and just the *thought* of peeling it makes you tired, a vegetable spiral slicer is a handy gadget to have. You'll be cranking out mounds of curly noodles in mere minutes.

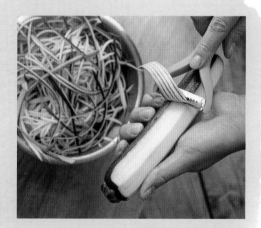

BOTTOM-OF-THE-BOX
BIBIMBAP

MAKES 4 SERVINGS

for the sauce

¼ cup (88 g) gochujang (Korean red pepper paste, see sidebar)

2 tablespoons (28 ml) hot water

1 tablespoon (15 ml) rice vinegar

1 tablespoon (15 ml) sesame oil

1 tablespoon (8 g) toasted sesame seeds

2 garlic cloves, minced

for the rice

2 cups (390 g) uncooked white rice, rinsed

4 cups (946 ml) water

for the bibimbap

8 cups (240 g) packed spinach

1 tablespoon (15 ml) sesame oil

1 teaspoon toasted sesame seeds

3 tablespoons (45 ml) olive oil, divided

2 zucchini, thinly sliced

2 carrots, cut into 3-inch (7.5 cm) matchsticks

1 cup (85 g) packed pea sprouts

4 eggs

Bibimbap—or b-bap, as I like to call it—is a classic Korean rice bowl. You may have seen it served in Korean restaurants in a piping-hot, heavy stone bowl with a raw egg cracked into it. That version, dolsot bibimbap, uses a stone bowl so hot it sizzles everything that touches it, including the rice and the egg. I don't get that fancy here, but I do serve my b-bap with the signature red sauce that coats all the rice, the vegetables, the egg—everything, really, since the word *bibimbap* means "mixing rice," and the only proper way to eat it is to mix it all up with chopsticks before you dive in!

I keep my b-bap simple and use whatever I can find in the kitchen. This is a good way to use up excess squash from your garden, leftover carrots or greens, an opened bag of sprouts, a mushroom or two—anything that might go bad soon. I like a combination of seasoned spinach, sautéed vegetables, and fresh sprouts with a sunny-side-up egg. The runny yolk is essential for bringing all the ingredients together in the bowl. If you plan ahead, I highly recommend making and adding some Quick-Pickled Sweet 'n Spicy Radish Pods (page 140) to the bowl, too. (Try it with radishes if you don't have radish pods.)

To make the sauce, combine all of its ingredients in a small bowl and set aside.

To make the rice, bring the rice and the water to a boil in a medium saucepan. Reduce the heat, cover, and simmer until the water is absorbed and the rice is cooked through, about 25 minutes.

Meanwhile, bring another medium saucepan of salted water to a boil. Blanch the spinach for 1 minute, then drain and thoroughly squeeze out all the liquid. Gather the spinach into a clump and coarsely chop. Transfer the spinach to a large platter, then drizzle the sesame oil and scatter the sesame seeds on top.

Heat a large skillet over medium-high heat and add 1 tablespoon (15 ml) of the olive oil. Stir in the zucchini and cook until tender, about 3 minutes. Transfer the zucchini to the same platter as the spinach.

In the same skillet over medium-high heat, add 1 tablespoon (15 ml) of the olive oil, stir in the carrots, and cook for about 5 minutes until tender. Transfer the carrots to the platter.

Divide the rice among four bowls and arrange the spinach, zucchini, carrots, and pea sprouts on top of the rice.

Reheat the same skillet over medium-high heat and add the remaining 1 tablespoon (15 ml) of olive oil, swirling it around to coat the surface. Crack the eggs into the skillet, making sure the whites don't run into each other. (If your skillet won't fit all the eggs at once, fry them in batches.) Cover and reduce the heat to low. Cook until the whites are set and the edges start to brown, about 5 minutes. Top each bowl with a fried egg and serve with a few spoonfuls of sauce to taste.

A PASTE IS NOT A POWDER

Gochujang is a thick and concentrated red pepper paste, typically sold in small tubs or jars at the Korean market. It's rich and spicy with undertones of sweet, a fermented product made from red chiles, glutinous rice, and fermented soybeans—not to be confused with gochugaru, the red pepper powder used in kimchi (see page 165). The two are not interchangeable, as they have distinctive flavor profiles.

DRUNKEN PUMPKIN CHILI

**MAKES 10 TO 12
SERVINGS**

2 tablespoons (28 ml) olive oil

3 pounds boneless pork shoulder, trimmed and cut into 1-inch (2.5 cm) pieces

2 yellow onions, chopped

6 garlic cloves, chopped

2 bell peppers, chopped

2 Anaheim chiles, chopped

1 poblano chile, chopped

1 habanero chile, minced

3 chipotle chiles canned in adobo sauce, chopped

3 cups (348 g) diced pumpkin

2 cups (308 g) sweet corn kernels

2 cans (15 ounces, or 425 g each) of kidney beans, drained and rinsed

1 can (28 ounces, or 785 g) of crushed tomatoes, undrained

1 can (6 ounces, or 170 g) of tomato paste

1 bottle (12 ounces, or 355 ml) of dark beer

1 cup (235 ml) strong brewed black coffee

4 cups (946 ml) chicken broth

1 tablespoon (7 g) ground cumin

1 tablespoon (3 g) dried oregano

¼ cup (60 g) packed brown sugar

Sour cream for serving

Chopped scallions for serving

Shredded cheddar, Monterey Jack, or Colby cheese for serving

Don't be intimidated by the long list of ingredients here. This is a "dump it and forget it" kind of recipe, the perfect one-pot meal to please and feed a crowd when you're tucked away in a mountain cabin and everyone is starving after a day on the slopes. I guarantee that as soon as the first mouthwatering smells of slow-cooked pork and spicy, smoky chiles start wafting through the air, your friends will let you have first tracks the next morning.

Despite the name, neither the pumpkin nor the beer will get you drunk, but I do recommend a very dark beer (the darker the better) for added richness. Try a brown ale or a stout.

Heat a large pot over medium-high heat. Drizzle in the oil and add the pork shoulder, stirring periodically until the pork is pinkish-brown on all sides, about 10 minutes. Add the onions and garlic and cook for 2 to 3 minutes until the onions start to turn translucent. Add all the bell and chile peppers and cook for another 5 minutes. Add the pumpkin, corn, 1 can of kidney beans, crushed tomatoes, tomato paste, beer, coffee, and broth. Stir in the cumin, oregano, and brown sugar, then bring the pot to a boil.

Once the chili reaches a continuous boil, reduce the heat and simmer uncovered for 1 hour, giving it an occasional stir to make sure the ingredients are well combined. Add the remaining 1 can of kidney beans and continue to simmer for 1 hour more. Serve with a dollop of sour cream, a spoonful of chopped scallions, and a handful of shredded cheese on top.

CHILE PEPPER PRIMER

What I love most about this dish is the way you can personalize it to make your own signature chili. You can use another orange-fleshed winter squash in place of the pumpkin, simmer other cuts of meat, omit the corn, or add some carrots; but whatever you do, do not skimp on the chiles. After all, this is pumpkin *chili*, not pumpkin stew.

The variety of chile peppers used in this recipe offers a complexity in flavor you can't get from simply using chili powder, a common ingredient in many chili recipes. The store-bought powdered version is never as balanced or rich as real chiles are. First, we have a layer of sweetness from fresh Anaheim chiles, the mildest of the peppers. Then, we have a rich and fruity note from the poblano, heat

from the habanero, and smokiness from the chipotles in adobo sauce. If you're sensitive to spice, you can substitute a cayenne pepper (medium hot) or serrano pepper (least hot) for the habanero.

When you're out shopping, keep in mind that an Anaheim chile is sometimes called a New Mexico chile or California chile. The young pepper is green and gradually matures into red, at which point it's called a chile colorado. To make matters even more confusing, many American grocers sometimes mislabel poblano chiles as pasilla chiles, even though true pasillas are actually the dried form of chilaca chiles. If you are sourcing these chiles from a grocer, look for a fresh, wide, dark green pepper.

AUTUMN ACORN SQUASH
STUFFED WITH KALE, CRANBERRIES, AND WALNUTS

MAKES 4 SERVINGS

2 acorn squashes (1½ pounds, or 680 g each)

2 tablespoons (28 ml) olive oil, divided

A few pinches of kosher salt

A few grinds of black pepper

2 shallots, chopped

2 garlic cloves, chopped

1 cup (190 g) uncooked brown rice, rinsed

2¼ cups (535 ml) chicken broth

2 cups (134 g) chopped kale, ribs and stems removed

2 tablespoons (5 g) chopped fresh sage

2 tablespoons (28 ml) balsamic vinegar

¾ cup (90 g) dried cranberries

½ cup (60 g) toasted and chopped walnuts

¼ to ½ cup (38 to 75 g) crumbled feta cheese

As soon as the weather turns and mounds of winter squash start appearing at the market, I find every reason to turn on the oven, steam up the windows, and make a warming autumn meal that fills the house with all kinds of good smells. This recipe is one of those reasons!

It's not exactly a one-pot meal. It's more like an all-in-one meal: mouth-meltingly tender roasted squash heaped with herb-scented brown rice, silky kale, and cranberries and walnuts. A splash of balsamic cuts the sweetness, and a sprinkle of feta adds some sharpness. If I was making this for my more carnivorous husband, I might be inclined to bake a strip of bacon in the oven with the squash and then crumble it over the rice. That's the beauty of a stuffed squash: You can stuff it with just about anything.

Preheat the oven to 425°F (220°C, or gas mark 7).

Slice the squashes in half lengthwise, then scoop out the stringy centers and seeds until you get a nice hollow in each squash. Place them cut-sides up on a large rimmed baking sheet and drizzle 1 tablespoon (15 ml) of the oil over them. Sprinkle the salt and pepper on top and roast the squashes for about 40 minutes until the edges are browned and the flesh is easily pierced with a fork. When the squashes are done, keep them warm in the oven until they are ready to serve.

Meanwhile, heat a large saucepan over medium heat. Swirl in the remaining 1 tablespoon (15 ml) of oil and add the shallots and garlic. Cook until the shallots start to turn translucent, 2 to 3 minutes. Stir in the rice and cook for 30 seconds to bring out its fragrance. Add the broth and bring to a boil. Reduce the heat, cover, and simmer for about 40 minutes until most of the liquid is absorbed. About 15 minutes before the rice is done, add the kale and sage right on top and cover it again. There is no need to stir just yet, as the leaves will steam and wilt on the surface while the rice continues to cook underneath them.

When the rice is fully cooked, turn off the heat and let stand for 5 minutes. Uncover, then mix in the kale and sage and fluff with a fork. Stir in the vinegar, cranberries, and walnuts.

Remove the squashes from the oven. Heap the rice filling into each squash and top with crumbled feta.

COOK'S NOTE
If you don't have any acorn squash on hand, try the recipe with kabocha, red kuri, delicata, or another orange-fleshed winter squash.

HIGH-FIVES & HUGS

This book is a rebirth of my debut title from 2015, *The CSA Cookbook*, but credit still goes to the original folks who helped make it happen, as well as to the new team who believed it needed to be seen again. I am thrilled to have this book in your hands. Before this, it made its way through several hands, both literal and virtual, and involved a year-long collaboration with all the people who helped inspire, create, and critique the ideas, stories, and recipes on these pages. It's my pleasure to give them proper praise, high-fives, and hugs all around.

To the husband of the century, Will Taylor, my talented photographer, constant cheerleader, and official taste tester, whose beautiful work adorns these pages, thank you for believing in the project before it was ever inked, washing the dishes when I was too tired to even stand, and putting up with my months of reclusiveness while I was writing. Above all, thank you for capturing this incredible moment in our lives with such style. I wouldn't have done it with anyone else but you. I love you to the moon and back.

Much love to my daughters, Gemma Lumen and Ember Luna, for inspiring me daily to be a better person, to infuse love and creativity into everything I do, and to savor the beauty and minutiae of everyday life.

Cảm ơn rất nhiều to my family, who always kept me well fed (sometimes *too* well fed!). Endless love to Dad and Tony, for making sure there was a homemade meal on the table every night of the week, and to Mom, for always making me eat my vegetables. Everything I learned about life and food started at home.

To my editor at The Harvard Common Press, Thom O'Hearn, whose serendipitous email arrived at the exact moment the idea for this book started bouncing around in my head. Thank you for seeing the potential in what I was doing with *Garden Betty* and for guiding this book to fruition (a second time).

Immense thanks to my mother-in-law, Sally Taylor, for letting us invade her kitchen, raid her cupboards, occupy her fridge, and stuff her silly with food during a week of photo shoots at Chamizal. It warms my heart to be able to preserve a bit of the family history in this book.

To my siblings-in-law, Richard and Tracy Taylor, for letting us take over their handsome homes in Point Reyes and San Francisco for a week of shooting. Being in such idyllic places made the long days seem more like play than work. Thank you.

I owe a round of beers to my enthusiastic and intrepid recipe testers, whose invaluable feedback and suggestions elevated these recipes to something worthy of my readers: ErinBlythe Sanders, Erin Murtaugh, Stephen Le, Lee Anne and Alex Dombroski, Tracy Grubbs and Richard Taylor, Austin and Tomoko Bryant, Sandy Cole and Bo Cross, Jason Knott and Jon Lucchese, Jennifer Sankary and Alan Falgout, Amanda and Jebb Stewart, Stanley Law, Lisa La, and Aaron Takahashi.

To Patricia and Norman Tuck, who unwittingly initiated and encouraged my current path. Without them, my blog and this book would not exist. Thank you for building the beautiful garden that inspired my cooking, and for having the foresight, decades ago, to create the sanctuary I was proud to call home.

And last but certainly not least, a huge high-five and big bear hug to every person who follows the *Garden Betty* blog. I never expected the site to take off as it did and to introduce me to a whole community of people who are as passionate about living well and eating well as I am. Thank you for sharing your stories, comments, hearts, and minds with me over the years. I'm beyond blessed.

RESOURCES

Most of the ingredients in this book can be found at your local grocer or ethnic market, but if you have trouble sourcing certain items locally, these are my go-to vendors online (some of which have brick-and-mortar locations).

AZURE STANDARD
www.azurestandard.com
Olive oil, sunflower oil, nuts, tahini, and sesame seeds (white, brown, and black)

BAKER CREEK HEIRLOOM SEEDS
www.rareseeds.com
Thousands of heirloom seeds for growing your vegetables, fruits, flowers, and herbs

HMART
www.hmart.com
Gochugaru and gochujang

LA TOURANGELLE
www.latourangelle.com
Artisan seed and nut oils

LOCALHARVEST
www.localharvest.org
Find a farm, farmers' market, or CSA (community supported agriculture) program near you

TOKYO CENTRAL
www.tokyocentral.com
Rice vinegar, Asian (toasted) sesame oil, and dried soba (buckwheat) noodles

THE MEADOW
www.themeadow.com
Artisan salt

MEXGROCER
www.mexgrocer.com
Canned chipotles in adobo sauce, Cotija cheese, crema Mexicana, and pepitas

PENZEYS SPICES
www.penzeys.com
Spices and soup bases

RED BOAT FISH SAUCE
www.redboatfishsauce.com
Pure first-press Vietnamese fish sauce

WAN JA SHAN
www.wanjashan.com
Organic soy sauce and organic gluten-free tamari

ABOUT THE AUTHOR

After more than a decade of growing and preserving her own food, raising chickens in urban backyards, and trying to craft a more sustainable and simple life (which brought her to the Pacific Northwest from the California coast), **Linda Ly** finally found a place to put down roots. She and her photographer husband, Will, are embarking on the grand adventure of building their first home together in Bend, Oregon, with dreams of turning their bare land into a modern homestead, food forest, and learning playground for their two daughters.

Follow along at www.gardenbetty.com, where Linda will be sharing their journey, as well as more recipes, projects, and stories, on her award-winning blog. You can also find her on social media as @gardenbetty.

ABOUT THE PHOTOGRAPHER

From as early as he can remember, **Will Taylor** was sneaking off with his family's cameras and snapping the people and places around him. But it was a split-second moment that he captured of Lake Tahoe, on an especially mesmerizing evening, that ignited his passion for storytelling and motivated him to pursue a career in photography. Decades later, his work in landscapes, lifestyle, fashion, and food has taken him around the world and appeared in dozens of publications (including his wife Linda's cookbooks).

Keep up with Will at www.instagram.com/willtaylorphotography, where you'll find him and his favorite subjects—his family and friends—camping, paddling, snowboarding, mountain biking, and living the #bendlife in Central Oregon.

INDEX